SOLAR ENERGY for the NORTHEAST

SOLAR ENERGY for the NORTHEAST

MARK UHRAN

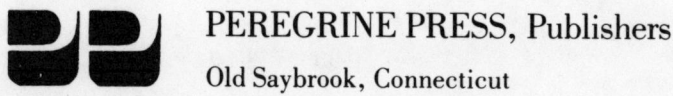

PEREGRINE PRESS, Publishers
Old Saybrook, Connecticut

Cover design by Marta Cone

Book design by Hildebrandt Associates

Copyright © 1980 by Mark Uhran. All rights reserved. No portion of this book may be reproduced in any form without written permission of the publisher, except by a reviewer who may quote brief passages in a review.

Manufactured in the United States of America

First Printing

ISBN 0-933614-04-7

Table of Contents

Introduction	3
Preliminary Considerations	5
The Alternatives	25
Solar Collection	39
Solar Storage	81
Solar Distribution	107
Solar Systems	117
Appendices	
Manufacturers of Solar Equipment	145
Additional Reading	148
Glossary of Terms	149
Index	157

INTRODUCTION

Energy use has risen dramatically into the forefront of our day to day lives, presenting itself as a threat to the warmth and quiet security of family and home. It represents a new challenge which demands attention and innovation if we are to preserve the life style we have so innocently enjoyed over the past few decades. Energy costs have climbed at an unprecedented and alarming rate; we now realize that a barrel of oil or a tankful of gasoline are far more valuable to us than we could have ever anticipated.

The search for clean and abundant new energy sources has led us roundabout to the oldest and most dependable resource the planet has: the sun. The sun has always, indirectly, provided for all of our needs, and in the past decade it has proved, undeniably, that it can continue to do so. We can use both simple and complex strategies to allow the sun to heat our homes, nourish our crops year-round and ultimately, provide our electricity. If we choose to accept its beneficent glow.

The gamut of solar technology runs from such simple methods as using large glass windows on the south walls of homes to the sophisticated space age engineering of orbiting power satellites which could capture and beam energy down to the planet's surface by microwave relay. Admittedly, the sophisticated methods are as yet largely untried. However, the basic technologies, those which we may apply to our own homes, are proven and practical. In some cases these methods are inexpensive and easily implemented, requiring only that the user take the time to search out and understand the ways of the sun.

At this time, there is no question as to the most efficient and economical application of solar energy: heating water. Hot water, needed for a wide variety of domestic uses, is the largest consumer of energy in the home aside from space heating. In the Northeast, solar hot water installations are quickly becoming the most favored means of conservation due to their marked success. With the aid of federal and state programs, literally thousands of solar hot water systems have been installed in the Northeast, serving as monuments to the pioneers of energy independence.

The workings of these systems are not incomprehensible. In fact, the basic concepts are quite ordinary and require only a new examination of the fundamentals involved in heat transfer. By looking closely at the principles of solar collection, solar storage, and solar distribution, we can begin to see the ease with which the heat of the sun's rays can be captured, contained, and moved to where we need it most. Once this understanding is achieved, the variety of ways in which we can apply it is limited only by the imagination of the innovator.

The purpose of *Solar Energy for the Northeast* is to aid the concerned energy consumer in the decisive action necessary to overcome the obstacles of expensive and increasingly unobtainable fossil fuels. In addition, it encourages the rediscovery of our most faithful and dedicated resource. As the cost of energy rises, so also the sun will continue to rise, and with each morning its significance as an energy provider becomes only more profound.

Mark Uhran

1: PRELIMINARY CONSIDERATIONS

The intensity of solar radiation varies depending on many topographical and vegetative features of a particular area. For instance, if you live on the north side of a mountain you need not read any further; solar energy is not for you. However, in many cases builders and architects have taken advantage of the sun by orienting homes with large expanses facing south. This provides the most basic form of solar heating and is termed passive solar design.

PASSIVE DESIGN

When planning a new home, or improving an existing home, the passive element should be considered first as it is a prerequisite to an overall energy efficient design. The most common passive systems are easy to build, and inexpensive to install. They are also difficult to appreciate unless artfully incorporated into the home's facade and interior. Typically, passive design is associated with large glass areas on the south side of the building. In this manner the sunlight reaches inside and warms the house during the day. If you wish to retain this energy for nighttime use, a suitable storage media must be provided. The storage media must have a high thermal mass; in other words, the ability to contain an 8-10 hour supply of thermal energy which will last until the sun comes up the next morning. This often requires large spaces for the storage material. Depending on the size of the home, one to two thousand gallons of water or 500 to 1000 cubic feet of rock may be in order. Locating this thermal mass, in a manner which is congruent with the home, presents a difficult design task. The alternatives are many and several interesting new forms in building design have emerged as a result.

Figure 1, Passive Design Alternatives, illustrates four basic approaches. There is no way to transfer the heat within the structure except by natural means: convection, conduction, and radiation. Nor is there any provision for long term heat storage or automatic control of the building temperature. However, these systems can provide the home with large amounts of "free" solar energy.

Passive solar systems perform two extremely important functions.
1. To increase overall heat gain to the building.
2. To control and limit re-radiation (conduction and convection heat losses from the building).

The first function is achieved by properly orienting the building. For maximum energy collection, the main axis of the home should be located east-west with the greatest expanse of window area facing south or up to 20° west of south.

The second function is controlled by careful and thorough insulation of the floor, walls and ceiling of the living space. Thermal curtains on all windows are a must. Examples of several low cost home-fashioned curtains are given in Chapter Six, Systems. Even a well built thermopane window will lose up to twice as much heat if not protected by an insulating curtain or equivalent thermal barrier. Provision should be made so that this curtain fits snugly around the window perimeter. Often the cold space between the curtain and glass will have a slightly lower pressure, due to the cooling air, and cause the curtain to be drawn in slightly, forming a seal on all sides.

1: Preliminary Considerations

PASSIVE DESIGN ALTERNATIVES

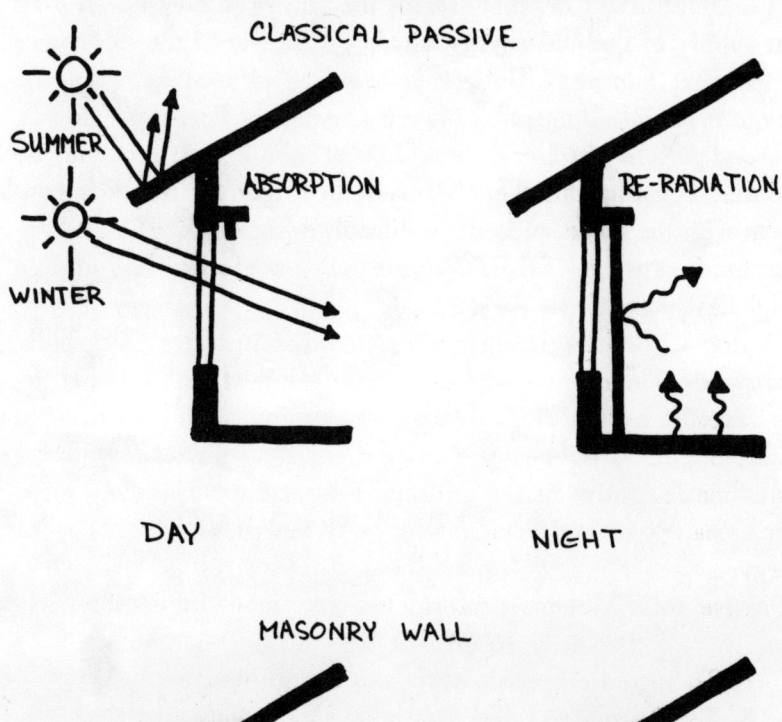
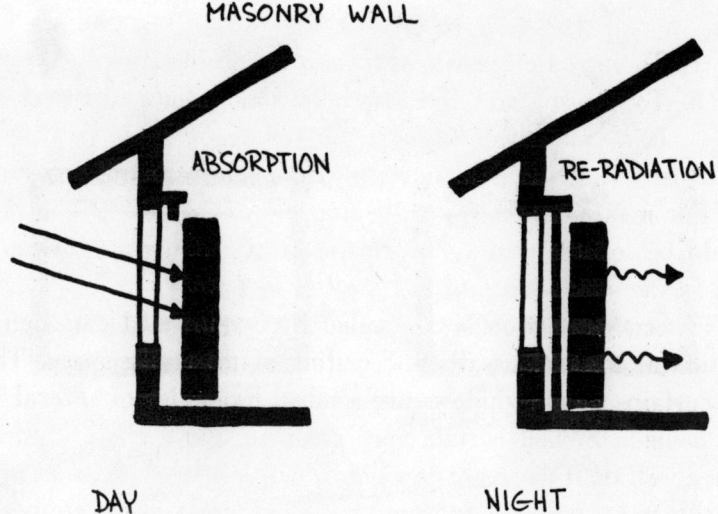

FIGURE 1-A

PASSIVE DESIGN ALTERNATIVES

DRUMWALL

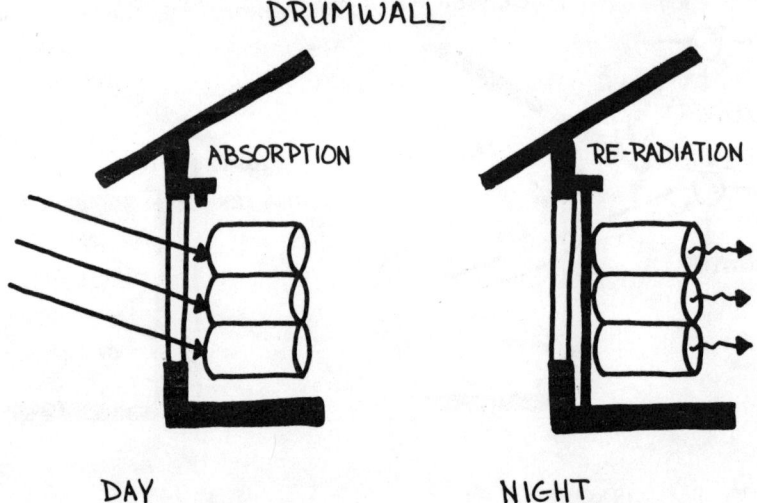

ROOF POND

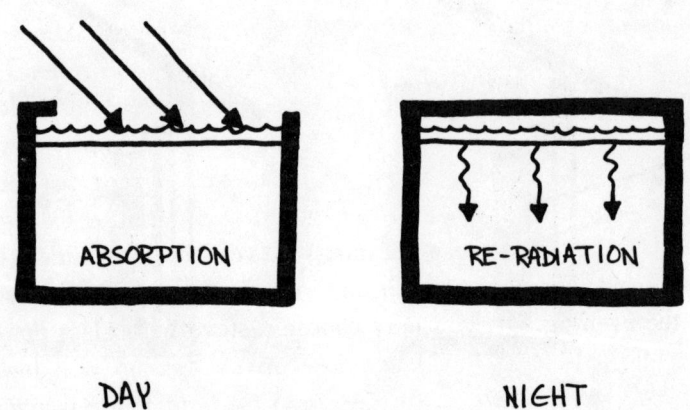

FIGURE 1-B

1: Preliminary Considerations

Determine which areas of the home will require a comfortable temperature range and then isolate these areas from the rest of the house. Generally this means extra insulation in the attic floor and basement ceiling. In addition, you may install furnace ducts which are easily opened and closed so that bedrooms are not heated during the day and living or family rooms are not heated at night. Admittedly, it's a bit tiresome to have to run around a home opening and closing vents but not nearly as tiresome as rising fuel costs.

When conventional heating systems are installed, insist that all hot air ducts be located on interior walls away from the building's outer skin. The worst possible place for a heating vent is on an outside wall under a window. All the energy immediately rises to displace cooler air adjacent to the window. It then cools, due to the large temperature difference between the two air masses on either side of the glass.

There are other areas of your house where hot air is deliberately blown outside even during the winter. These include the clothes dryer, kitchen and bathrom vent fans, and the fireplace. The first two can be corrected by installing charcoal filters and venting the air indoors. If kitchen or bathroom odors are persistent then vent the air into the basement but, by all means, keep it inside. The fireplace can be modified to accept a wood burning stove. These stoves have everything in their favor; the wood will burn slower and less heat will escape through the chimney with a properly set damper. For those of you who insist on being able to see your fire while it's burning, there are many stoves now available with large front-opening doors which can remain open and be closed later in the evening. Or, you may choose a stove with glass doors.

During the winter season leave the furnace fan on at a lower speed if you have a forced air system. This is called continuous air circulation and will prevent air from stratifying into layers where the floor is warmer than the ceiling. This even heating will cause your thermostat to react more consistently and prevent the furnace from firing due to minor temperature fluctuations. In addition it prevents the loss of hot air up the furnace flue during the time when the burner is coming on and going off. The additional

Solar Energy for the Northeast

cost of running the electric fan motor continuously is minor compared to the fuel savings which result.

All furnace ductwork in the basement area should be insulated unless you plan on heating the basement. Also, any areas of the foundation wall which are above ground level are energy losers and should be covered with foam or fiberglass batts to minimize losses. A more significant improvement is to backfill against the outside of the house and create an earth berm which surrounds the building perimeter as high up as practical. If you are not intrigued by this approach, then at least plant a dense foliage around the perimeter to act as a windbreak.

Having accomplished all of the previous minor improvements, you can be assured that you now have an energy conservative passive solar home. These items, although individually small, may save you between 10 and 20% on your annual fuel requirements.

Be aware of how energy is generated in your home, where it comes from, and how it moves from place to place. There are certainly many preventive measures which are not listed here but may apply to your home. A casual walk through your house on a cold windy day will point out small areas where cold air can be felt coming through cracks around windows, walls and electric outlets. A caulk gun with a silicone cartridge is the best means to deal with these small annoyances.

Be energy creative in day-to-day activities. At $1.00 per gallon for fuel oil or 10¢ per kilowatt hour for electricity, energy creativity becomes profitable. There are many not so practical but humorous means to be energy creative also. For instance, after bathing, don't run the water down the drain. If a bathtub holds 50 gallons of water at 88°F then there are 8,300 Btus of thermal energy hidden there. Allow the water to cool to 68°F, or room temperature, before draining, and let that energy heat the home instead of the waste water piping. This in turn will leave a ring around the tub and your body will expend another 500 Btus of energy to clean it off. (Add a wetting agent such as detergent or bubble bath if you are not that ambitious.)

1: Preliminary Considerations

If you are expecting a particularly cold evening, invite people into your home to heat it. An average person produces approximately 400 Btus per hour while sitting or resting. If your home has a heating load of 24,000 Btus per hour, 60 people will heat it quite nicely. (If you can get them dancing, 20 will do equally well.) Needless to say, energy in one form or another surrounds us inconspicuously in our day-to-day lives. Understanding where that energy is and what it means to us is the first step in becoming energy self reliant.

SITE EVALUATION

There are numerous charts, graphs and tables available for estimating the amount of solar energy at a specific site. The most accurate and dependable are those references provided by the United States Meteorological Service and ASHRAE (the American Society of Heating, Refrigerating and Air Conditioning Engineers). Unfortunately, data is not available for every location in New England. This becomes a particular problem because microclimates, with unusually persistent conditions, are quite common in the Northeast. Generally a data point can be located within reasonable proximity to the site, however, and local topography should be considered as well. For example, areas of higher altitude and north facing slopes pose problems. Heating requirements increase for a mountaintop home due to wind chill and the cooler elevation. Because the sun doesn't shine directly on a north facing slope, only diffuse radiation is available.

In addition, the weather data is based on statistical averages and is an estimate only. Normals, means and extremes are all part of the overall picture of New England climate. Year to year changes in total hours of sunshine or mean sky cover may appear to differ significantly, though a 20-year average gives a fairly reliable set of design conditions from which to work.

Figure 2, Winter Degree Days and Design Temperature, shows the normal extreme low temperature for a given area as well as the average heating requirements listed as heating degree days. A heating degree day is a good measure to determine relative requirements between two separate areas. A heating degree day occurs if the outdoor temperature is one degree below room temperature (standard practice is a room temperature of 65°F) for one day. The heating load is represented as one degree day. Consequently, if the average outside temperature is 32° for one day, 33 degree days (65 − 32 = 33) result.

In New England, heating requirements vary between five and nine thousand heating degree days per season. On close examination you will notice that the heating requirements do not necessarily increase as you move north. For example, the state of Vermont shows between 8000 and 8800 degree days per year whereas its northern neighbor, the state of Maine, has 7600 to 8000 degree days for the same period. In like manner, the winter design temperature is approximately 4°F lower in Vermont. The reason lies in the fact that Maine receives more total hours of sunshine.

This brings us to the United States Weather Service Data provided in Figure 3, Mean Total Number of Hours of Sunshine. This table indicates that Maine receives, on the average, 303 more hours of sunshine than Vermont. Translated into units of energy at the site this is equivalent to slightly over 90,000 Btus per square foot of solar collector.

The Northeast can expect between two and three thousand hours of sunshine during a one-year period. On a daily basis this amounts to between 5 and 8 hours per day. Not all of this energy is direct radiation—a significant portion is termed scattered or diffuse radiation. A flatplate solar collector captures diffuse radiation as well as direct and thereby qualifies itself as the most efficient collector type to be used in the Northeast. This subject will be examined again more closely in Chapter 3, Solar Collection.

For most locations in the Northeast, the total number of hours of sunshine is adequate to install a solar collection system and to be assured that it will remain cost-effective. Obviously, the areas of high cloud cover represent the less effective sites for solar energy use. These regions are often found to the east of large bodies of

1: Preliminary Considerations

DESIGN TEMPERATURE AND HEATING DEGREE DAYS

	ASHRAE DESIGN TEMP.	HEATING DEGREE DAYS		ASHRAE DESIGN TEMP.	HEATING DEGREE DAYS
CONNECTICUT			**NEW HAMPSHIRE**		
BRIDGEPORT	4°F	5,600	CONCORD	-11°F	7,400
DANBURY		6,000	DOVER		7,200
GREENWICH		5,400	KEENE	-12°F	7,400
HARTFORD	1°F	6,200	LACONIA		7,800
MERIDEN		6,000	MANCHESTER	-5°F	7,200
NEW BRITAIN		6,000	NASHUA		7,000
NEW HAVEN	5°F	5,800	PORTSMOUTH	-2°F	7,200
NORWALK	0°F	5,400	**NEW YORK**		
STAMFORD		5,400	ALBANY	-5°F	6,800
MAINE			BINGHAMTON	-2°F	7,200
BANGOR	-8°F	8,000	BUFFALO	3°F	7,000
LEWISTON	-8°F	7,800	ELMIRA	1°F	6,400
PORTLAND	-5°F	7,600	GLENS FALLS	-11°F	7,200
MASSACHUSETTS			NEW YORK	12°F	5,000
BOSTON	6°F	5,800	NIAGARA FALLS	4°F	7,000
CAPE COD	5°F	6,000	POUGHKEEPSIE	-1°F	6,200
FALL RIVER	-3°F	5,800	ROCHESTER	2°F	6,800
LAWRENCE	-1°F	6,800	ROME	-7°F	7,400
LOWELL	-1°F	6,800	SCHENECTADY	-5°F	6,800
PITTSFIELD	-5°F	7,600	SYRACUSE	2°F	6,800
SPRINGFIELD	-3°F	6,600	WATERTOWN	-14°F	7,200
WORCESTER	-3°F	7,000	WHITE PLAINS		5,600
VERMONT			**RHODE ISLAND**		
BURLINGTON	-12°F	8,200	NEWPORT	5°F	5,800
MONTPELIER		8,800	PROVIDENCE	6°F	6,000
RUTLAND	-12°F	8,000	WESTERLY		5,800

FIGURE 2

Solar Energy for the Northeast

MEAN TOTAL NUMBER OF HOURS OF SUNSHINE

	JAN	FEB	MAR	APR	MAY	JUN	JUL	AUG	SEP	OCT	NOV	DEC	YEAR
CONNECTICUT													
HARTFORD	141	166	206	223	267	285	299	268	220	193	137	136	2,541
NEW HAVEN	155	178	215	234	274	291	309	284	238	215	157	154	2,704
MAINE													
EASTPORT	133	151	196	201	245	248	275	260	205	175	105	115	2,309
PORTLAND	155	174	213	226	268	286	312	294	229	202	146	148	2,653
MASSACHUSETTS													
BOSTON	148	168	212	222	263	283	300	280	232	207	152	148	2,615
NANTUCKET	128	156	214	227	278	284	291	279	242	208	149	129	2,585
NEW HAMPSHIRE													
CONCORD	136	153	192	196	229	261	286	260	214	179	122	126	2,354
MT. WASHINGTON	94	98	133	141	162	145	150	143	139	159	88	87	1,540
NEW YORK													
ALBANY	125	151	194	213	266	301	317	286	224	194	115	112	2,496
BINGHAMTON	94	119	151	170	226	256	286	230	184	158	92	79	2,025
BUFFALO	110	125	180	212	274	319	338	297	231	183	97	84	2,458
NEW YORK	154	171	213	237	268	289	302	271	235	213	169	155	2,677
ROCHESTER	93	123	172	209	274	314	333	294	224	173	97	86	2,392
SYRACUSE	87	115	165	197	261	295	316	276	211	163	81	74	2,241
RHODE ISLAND													
PROVIDENCE	145	168	211	221	271	285	292	267	226	207	153	143	2,589
VERMONT													
BURLINGTON	103	127	184	185	244	270	291	266	199	152	77	80	2,178

SOURCE: U.S. METEOROLOGICAL DATA SERVICE

FIGURE 3

1: Preliminary Considerations

water. There is a simple reason for this phenomenon: water bodies are often warmer than the surrounding area, causing the air masses that move over them to cool as they reach land and lead to cloud formation. These clouds in turn cut down the solar transmissiveness in the area. A well known example of this situation occurs in western New York State where Lake Erie and Lake Ontario cause massive cloud formation over a substantial portion of the region. By examining the table on Hours of Sunshine it is shown that this is indeed the effect, as both Binghamton and Syracuse are represented by less than 2,200 hours of sunlight per year. Lake Champlain in Vermont has a similar effect on the northern regions of this state and results in 2,178 hours of sunshine in the Burlington area.

To examine the degree of cloudiness which prevails in a particular area refer to Figure 4, Mean Sky Cover, Sunrise to Sunset. You can see that each value is chosen from a range of 1 to 10 where 10 is a completely overcast sky. A value such as 6.8 can be translated to mean that on an average day 68% of the sky is cloudy. In New England, these values range from 7.1 in Caribou, Maine to 5.2 in Concord, New Hampshire.

It is also interesting to note that the difference between the summer and winter months is only about 5-10%. There is a common misunderstanding that because the winter season is colder, there is not a sufficient supply of radiant energy to warrant a solar collector system. In actuality, the sun shining on a winter day is as intense as the sun on a summer day, only the air temperature is lower. The lower air temperature will have an effect on collector operation efficiency, but this may be overcome significantly with proper collector design. Areas which are typified by cold clear days present the ideal conditions for winter solar energy collection. The demand for heating during the nighttime may be satisfied entirely by the sunlight collected on a clear 20°F day. This can not be accomplished by a simple box type air collector unless special attention has been given to reducing heat losses from the unit. Many other problems come into play also

15

Solar Energy for the Northeast

MEAN SKY COVER, SUNRISE TO SUNSET

	JAN	FEB	MAR	APR	MAY	JUN	JUL	AUG	SEP	OCT	NOV	DEC	AVG
CONNECTICUT													
HARTFORD	6.2	5.7	5.8	6.1	6.0	5.8	5.7	5.6	5.4	5.3	6.2	6.2	5.8
MAINE													
CARIBOU	7.2	6.8	6.8	7.3	7.4	7.5	7.0	6.8	6.6	6.8	8.1	7.3	7.1
EASTPORT	6.9	6.5	6.6	6.7	6.9	6.9	6.5	6.3	6.3	6.6	7.6	7.2	6.8
MASSACHUSETTS													
BOSTON	6.0	6.6	5.6	5.8	5.8	5.7	5.6	5.3	5.1	5.3	6.0	6.0	5.7
NEW YORK													
ALBANY	6.6	6.0	5.9	5.8	5.6	5.3	5.1	5.0	5.0	5.4	6.4	6.7	5.7
BINGHAMTON	7.5	7.2	7.1	6.9	6.5	6.2	6.1	6.0	6.0	6.3	7.5	7.9	6.8
BUFFALO	8.1	7.5	7.0	6.5	6.0	6.5	5.2	5.4	5.6	6.2	7.7	8.2	6.6
CANTON	6.9	6.3	6.0	6.1	5.7	5.0	4.8	5.0	5.5	6.1	7.5	7.4	6.0
NEW YORK	6.2	5.8	5.9	5.2	5.8	5.7	5.6	5.2	5.0	5.0	5.8	6.0	5.7
SYRACUSE	7.9	7.5	7.1	6.6	6.0	5.7	5.4	5.6	5.8	6.3	7.7	8.1	6.6
RHODE ISLAND													
BLOCK ISLAND	6.0	5.5	5.5	5.7	5.7	5.3	5.2	5.0	4.8	4.6	5.7	5.9	5.4
PROVIDENCE	5.9	5.4	5.4	5.7	5.6	5.4	5.4	5.1	5.0	4.7	5.5	5.7	5.4
VERMONT													
BURLINGTON	7.2	6.9	6.9	6.7	6.7	6.1	5.8	5.7	6.0	6.5	7.9	7.8	6.7
NEW HAMPSHIRE													
CONCORD	5.6	5.0	5.0	5.2	5.2	4.8	4.7	4.8	5.0	5.3	6.0	5.9	5.2

SOURCE: U.S. METEOROLOGICAL DATA SERVICE

FIGURE 4

1: Preliminary Considerations

but these can be resolved through various design improvements to be discussed later.

Having determined the mean cloud cover and total number of hours of sunshine available, you have a fairly reliable picture of the solar availability at your site. You must now choose a location in which the solar collectors will not be shaded by surrounding vegetation. Deciduous (trees with leaves) vegetation must be low enough that the sun will pass over the top on a summer's day. During the winter these trees will lose their leaves and only a minor portion of the sunshine will be lost, due to shading by trunk and branches. Coniferous (trees with needles) vegetation will remain opaque to sunlight all year round and therefore limit the use of collectors unless precautions are taken.

The best way to determine the sun's height in the sky (altitude) is to go outside and observe it over an eight-hour period. This is quick and easy compared to reading tables of latitude, hour angle and declination and converting them to your site. There are two simple rules to remember:

1. The sun is higher in the sky in the summer and low in the sky in the winter.
2. The sun rises south of due east in winter and north of east in the summer.

Based on this information you can determine whether or not vegetation will be a limiting factor. In some cases, trees may be topped off if the shading effect is significant. Another alternative is to install the collectors at the highest possible point—the roof.

Roof mounted solar collectors are the norm for two reasons; first, the shading effect, and second, the architectural convenience. A flat solar collector will blend in quite nicely with the roof lines of a home and often create the visual impression of sklights. By no means is it necessary, however, to always locate collectors on the roof. If the shading effect at ground level is not great, the units may be placed at ground level. If you are installing a home built solar system or an inexpensively pre-manufactured system this is important, because the collectors require maintenance. Replacing collector covers and repairing watertight seals is a much less irritating project if your feet are on the ground. Maintenance free collectors are available and you can expect to pay a higher price. If you are interested in a durable and reliable system which you can put on the roof and forget about, choose your manufacturer carefully and examine the collector thoroughly before buying. In the pages that follow, you will find the information necessary to make a sound choice.

Ground mounted collectors have the added feature of being compatible with reflector systems. A simple reflector plate of polished aluminum or stainless steel can improve solar pickup by as much as 15%. The drawback to reflectors is that, regardless of how much you pay for them, you still have to clean and polish them regularly or the reflective value drops off.

Once you have found that a particular site is suitable for locating collectors you will be interested in finding out exactly how much radiation will fall on the collecting surface. This can be determined by Figure 5, Solar Position and Insolation Values for Northern Latitudes. The table is provided by ASHRAE and lists the incoming solar radiation (Insolation) in British Thermal Units (Btus).

A Btu is the measure of energy required to raise one pound of water 1° Fahrenheit. For example, if you have a 60-gallon water tank and 1 gallon of water weighs 8.3 pounds then you have 498 lbs of water. To raise this water temperature by one degree requires 498 Btus. If you want to raise the water from 55° to 120°F then there is a 65°F difference and it will require 65 ✕ 498 or 32,370 Btus of thermal energy. Three solar collectors, each delivering 10,000 Btus per day, will handle this hot water requirement quite effectively.

How many Btus a solar collector will generate is dependent upon many factors to be discussed later in collector design. One overwhelming factor, however, is the amount of Btus falling on the collector. To determine this, first locate the table which corresponds to your latitude. Now pick the month of the year in which you are interested. Three values are given which may be defined as follows:

1: Preliminary Considerations

SOLAR POSITION AND INSOLATION VALUES 40°N

DATE	SOLAR TIME AM	SOLAR TIME PM	SOLAR POSITION ALT	SOLAR POSITION AZM	BTUH/SQ. FT. TOTAL INSOLATION NORMAL	HORIZ.	SOUTH FACING SURFACE 30	40	50
JAN 21	8	4	8.1	55.3	142	28	65	74	81
	9	3	16.8	44.0	239	83	155	171	182
	10	2	23.8	30.9	274	127	218	237	249
	11	1	28.4	16.0	289	154	257	277	290
	12		30.0	0.0	294	164	270	291	303
	SURFACE DAILY TOTALS				2182	948	1660	1810	1906
FEB 21	7	5	4.8	72.7	69	10	19	21	23
	8	4	15.4	62.2	224	73	114	122	126
	9	3	25.0	50.2	274	132	195	205	209
	10	2	32.8	35.9	295	178	256	267	271
	11	1	38.1	18.9	305	206	293	306	310
	12		40.0	0.0	308	216	306	319	323
	SURFACE DAILY TOTALS				2640	1414	2060	2162	2202
MAR 21	7	5	11.4	80.2	171	46	55	55	54
	8	4	22.5	69.6	250	114	140	141	138
	9	3	32.8	57.3	282	173	215	217	213
	10	2	41.6	41.9	297	218	273	276	271
	11	1	47.7	22.6	305	247	310	313	307
	12		50.0	0.0	307	257	322	326	320
	SURFACE DAILY TOTALS				2916	1852	2308	2330	2284
APR 21	6	6	7.4	98.9	89	20	11	8	7
	7	5	18.9	89.5	206	87	77	70	61
	8	4	30.3	79.3	252	152	153	145	133
	9	3	41.3	67.2	274	207	221	213	199
	10	2	51.2	51.4	286	250	275	267	252
	11	1	58.7	29.2	292	277	308	301	285
	12		61.6	0.0	293	287	320	313	296
	SURFACE DAILY TOTALS				3092	2274	2412	2320	2168
MAY 21	5	7	1.9	114.7	1	0	0	0	0
	6	6	12.7	105.6	144	49	25	15	14
	7	5	24.0	96.6	216	214	89	76	60
	8	4	35.4	87.2	250	175	158	144	125
	9	3	46.8	76.0	267	227	221	206	186
	10	2	57.5	60.9	277	267	270	255	233
	11	1	66.2	37.1	283	293	301	287	264
	12		70.0	0.0	284	301	312	297	274
	SURFACE DAILY TOTALS				3160	2552	2442	2264	2040
JUN 21	5	7	4.2	117.3	22	4	3	3	2
	6	6	14.8	108.4	155	60	30	18	17
	7	5	26.0	99.7	216	123	92	77	59
	8	4	37.4	90.7	246	182	159	142	121
	9	3	48.8	80.2	263	233	219	202	179
	10	2	59.8	65.8	272	272	266	248	224
	11	1	69.2	41.9	277	296	296	278	253
	12		73.5	0.0	279	304	306	289	263
	SURFACE DAILY TOTALS				3180	2648	2434	2224	1974

NOTE: 1) BASED ON DATA IN TABLE 1, pp 387 in 1972 ASHRAE HANDBOOK OF FUNDAMENTALS; GROUND REFLECTANCE; 1.0 CLEARNESS FACTOR.

FIGURE 5-1

19

Solar Energy for the Northeast

SOLAR POSITION AND INSOLATION VALUES 40°N

DATE	SOLAR TIME AM	SOLAR TIME PM	SOLAR POSITION ALT	SOLAR POSITION AZM	BTUH/SQ. FT. TOTAL INSOLATION NORMAL	HORIZ.	SOUTH FACING SURFACE 30	40	50
JUL 21	5	7	2.3	115.2	2	0	0	0	0
	6	6	13.1	106.1	138	50	26	17	15
	7	5	24.3	97.2	208	114	89	75	60
	8	4	35.8	87.8	241	174	157	142	124
	9	3	47.2	76.7	259	225	218	203	182
	10	2	57.9	61.7	269	265	266	251	229
	11	1	66.7	37.9	275	290	296	281	258
	12		70.6	0.0	276	298	307	292	269
	SURFACE DAILY TOTALS				3062	2534	2409	2230	2006
AUG 21	6	6	7.9	99.5	81	21	12	9	8
	7	5	19.3	90.0	191	87	76	69	60
	8	4	30.7	79.9	237	150	150	141	129
	9	3	41.8	67.9	260	205	216	207	193
	10	2	51.7	52.1	272	246	267	259	244
	11	1	59.3	29.7	278	273	300	292	276
	12		62.3	0.0	280	282	311	303	287
	SURFACE DAILY TOTALS				2916	2244	2354	2258	2104
SEP 21	7	5	11.4	80.2	149	43	51	51	49
	8	4	22.5	69.6	230	109	133	134	131
	9	3	32.8	57.3	263	167	206	208	203
	10	2	41.6	41.9	280	211	262	265	260
	11	1	47.7	22.6	287	239	298	301	295
	12		50.0	0.0	290	249	310	313	307
	SURFACE DAILY TOTALS				2708	1788	2210	2228	2182
OCT 21	7	5	4.5	72.3	48	7	14	15	17
	8	4	15.0	61.9	204	68	106	113	117
	9	3	24.5	49.8	257	126	185	195	200
	10	2	32.4	35.6	280	170	245	257	261
	11	1	37.6	18.7	291	199	283	295	299
	12		39.5	0.0	294	208	295	308	312
	SURFACE DAILY TOTALS				2454	1348	1962	2060	2098
NOV 21	8	4	8.2	55.4	136	28	63	72	78
	9	3	17.0	44.1	232	82	152	167	178
	10	2	24.0	31.0	268	126	215	233	245
	11	1	28.6	16.1	283	153	254	273	285
	12		30.2	0.0	288	163	267	287	298
	SURFACE DAILY TOTALS				2128	942	1636	1778	1870
DEC	8	4	5.5	53.0	89	14	39	45	50
	9	3	14.0	41.9	217	65	135	152	164
	10	2	20.7	29.4	261	107	200	221	235
	11	1	25.0	15.2	280	134	239	262	276
	12		26.6	0.0	285	143	253	275	290
	SURFACE DAILY TOTALS				1978	782	1480	1634	1740

FIGURE 5-2

1: Preliminary Considerations

SOLAR POSITION AND INSOLATION VALUES 48°N

DATE	SOLAR TIME AM	SOLAR TIME PM	SOLAR POSITION ALT	SOLAR POSITION AZM	BTUH/SQ. FT. TOTAL INSOLATION NORMAL	HORIZ.	SOUTH FACING SURFACE 38	48	58
JAN 21	8	4	3.5	54.6	37	4	17	19	21
	9	3	11.0	42.6	185	46	120	132	140
	10	2	16.9	29.4	239	83	190	206	216
	11	1	20.7	15.1	261	107	231	249	260
	12		22.0	0.0	267	115	245	264	275
	SURFACE DAILY TOTALS				1710	596	1360	1478	1550
FEB 21	7	5	2.4	72.2	12	1	3	4	4
	8	4	11.6	60.5	188	49	95	102	105
	9	3	19.7	47.7	251	100	178	187	191
	10	2	26.2	33.3	278	139	240	251	255
	11	1	30.5	17.2	290	165	278	290	294
	12		32.0	0.0	293	173	291	304	307
	SURFACE DAILY TOTALS				2330	1080	1880	1972	2024
MAR 21	7	5	10.0	78.7	153	37	49	49	47
	8	4	19.5	66.8	236	96	131	132	129
	9	3	28.2	53.4	270	147	205	207	203
	10	2	35.4	37.8	287	187	263	266	261
	11	1	40.3	19.8	295	212	300	303	297
	12		42.0	0.0	298	220	312	315	309
	SURFACE DAILY TOTALS				2780	1578	2208	2228	2182
APR 21	6	6	8.6	97.8	108	27	13	9	8
	7	5	18.6	86.7	205	85	76	69	59
	8	4	28.5	74.9	247	142	149	141	129
	9	3	37.8	61.2	268	191	216	208	194
	10	2	45.8	44.6	280	228	268	260	245
	11	1	51.5	24.0	286	252	301	294	278
	12		53.6	0.0	288	260	313	305	289
	SURFACE DAILY TOTALS				3076	2106	2358	2266	2114
MAY 21	5	7	5.2	114.3	41	9	4	4	4
	6	6	14.7	103.7	162	61	27	16	15
	7	5	24.6	93.0	219	118	89	75	60
	8	4	34.7	81.6	248	171	156	142	123
	9	3	44.3	68.3	264	217	217	202	182
	10	2	53.0	51.3	274	252	265	251	229
	11	1	59.5	28.6	279	274	296	281	258
	12		62.0	0.0	280	281	306	292	269
	SURFACE DAILY TOTALS				3254	2482	2418	2234	2010
JUN 21	5	7	7.9	116.5	77	21	9	9	8
	6	6	17.2	106.2	172	74	33	19	18
	7	5	27.0	95.8	220	129	93	77	59
	8	4	37.1	84.6	246	181	157	140	119
	9	3	46.9	71.6	261	225	216	198	175
	10	2	55.8	54.8	269	259	262	244	220
	11	1	62.7	31.2	274	280	291	273	248
	12		65.5	0.0	275	287	301	283	258
	SURFACE DAILY TOTALS				3312	2626	2420	2204	1950

NOTE: 1) BASED ON DATA IN TABLE 1, pp 387 in 1972 ASHRAE HANDBOOK OF FUNDAMENTALS; GROUND REFLECTANCE; 1.0 CLEARNESS FACTOR.

FIGURE 5-3

Solar Energy for the Northeast

SOLAR POSITION AND INSOLATION VALUES 48°N

DATE	SOLAR TIME AM	PM	SOLAR POSITION ALT	AZM	BTUH/SQ. FT. TOTAL INSOLATION ON SU NORMAL	HORIZ.	SOUTH FACING SURFACE 38	48	58
JUL 21	5	7	5.7	114.7	43	10	5	5	4
	6	6	15.2	104.1	156	62	28	18	16
	7	5	25.1	93.5	211	118	89	75	59
	8	4	35.1	82.1	240	171	154	140	121
	9	3	44.8	68.8	256	215	214	199	178
	10	2	53.5	51.9	266	250	261	246	224
	11	1	60.1	29.0	271	272	291	276	253
	12		62.6	0.0	272	279	301	286	263
	SURFACE DAILY TOTALS				3158	2474	2386	2200	1974
AUG 21	6	6	9.1	98.3	99	28	14	10	9
	7	5	19.1	87.2	190	85	75	67	58
	8	4	29.0	75.4	232	141	145	137	125
	9	3	38.4	61.8	254	189	210	201	187
	10	2	46.4	45.1	266	225	260	252	237
	11	1	52.2	24.3	272	248	293	285	268
	12		54.3	0.0	274	256	304	296	279
	SURFACE DAILY TOTALS				2898	2086	2300	2200	2046
SEP 21	7	5	10.0	78.7	131	35	44	44	43
	8	4	19.5	66.8	215	92	124	124	121
	9	3	28.2	53.4	251	142	196	197	193
	10	2	35.4	37.8	269	181	251	254	248
	11	1	40.3	19.8	278	205	287	289	284
	12		42.0	0.0	280	213	299	302	296
	SURFACE DAILY TOTALS				2568	1522	2102	2118	2070
OCT 21	7	5	2.0	71.9	4	0	1	1	1
	8	4	11.2	60.2	165	44	86	91	95
	9	3	19.3	47.4	233	94	167	176	180
	10	2	25.7	33.1	262	133	228	239	242
	11	1	30.0	17.1	274	157	266	277	281
	12		31.5	0.0	278	166	279	291	294
	SURFACE DAILY TOTALS				2154	1022	1774	1860	1890
NOV 21	8	4	3.6	54.7	36	5	17	19	21
	9	3	11.2	42.7	179	46	117	129	137
	10	2	17.1	29.5	233	83	186	202	212
	11	1	20.9	15.1	255	107	227	245	255
	12		22.2	0.0	261	115	241	259	270
	SURFACE DAILY TOTALS				1668	596	1336	1448	1518
DEC 21	9	3	8.0	40.9	140	27	87	98	105
	10	2	13.6	28.2	214	63	164	180	192
	11	1	17.3	14.4	242	86	207	226	239
	12		18.6	0.0	250	94	222	241	254
	SURFACE DAILY TOTALS				1444	446	1136	1250	1326

FIGURE 5-4

1: Preliminary Considerations

1. Normal: the amount of energy falling on a surface perpendicular to the sun's rays.
2. Horizontal: the amount of energy falling on a surface parallel to the earth's surface.
3. Angle with Horizontal: the amount of energy falling on various surfaces inclined to the earth's surface.

The value we are concerned with is the surface angle with horizontal. This corresponds to the inclination or pitch of the collector. For example, at 48° north latitude (Maine), a collector surface pitched 68° will receive approximately 1,578 Btus per square foot of surface on a January day. If you have a 20 square foot collector this amounts to 31,560 Btus per day. Remember, this is the amount reaching the collector, not the output. The efficiency of the collector will determine what proportion of this energy is captured and transferred to storage for later use.

These tables are excellent guides for predicting collector performance if you have the manufacturer's efficiency curves. If not, you can still determine relative performance between seasons. The tables also provide a convenient means of choosing the proper collector inclination angle. Unfortunately, data is not provided for every degree of inclination or latitude. You may interpolate the data where values are not given and still arrive at dependable estimates. It is important that you have a rough idea of the inclination you intend to use, to avoid running repetitious sets of calculations. A computer is convenient for this task but in most cases it is not available.

With data on cloud cover and total hours of sunshine you have been able to establish whether solar is worth pursuing at your site. With information on insolation you have narrowed this down to the specific amount of energy falling on your collectors. The remaining question is just how do you propose to capture this energy, store it, and put it to useful purposes. Many alternatives remain open at this point. Fortunately, the research already completed and implemented enables you to choose between these alternatives with confidence.

Solar Energy for the Northeast

SOLAR COLLECTOR TYPES

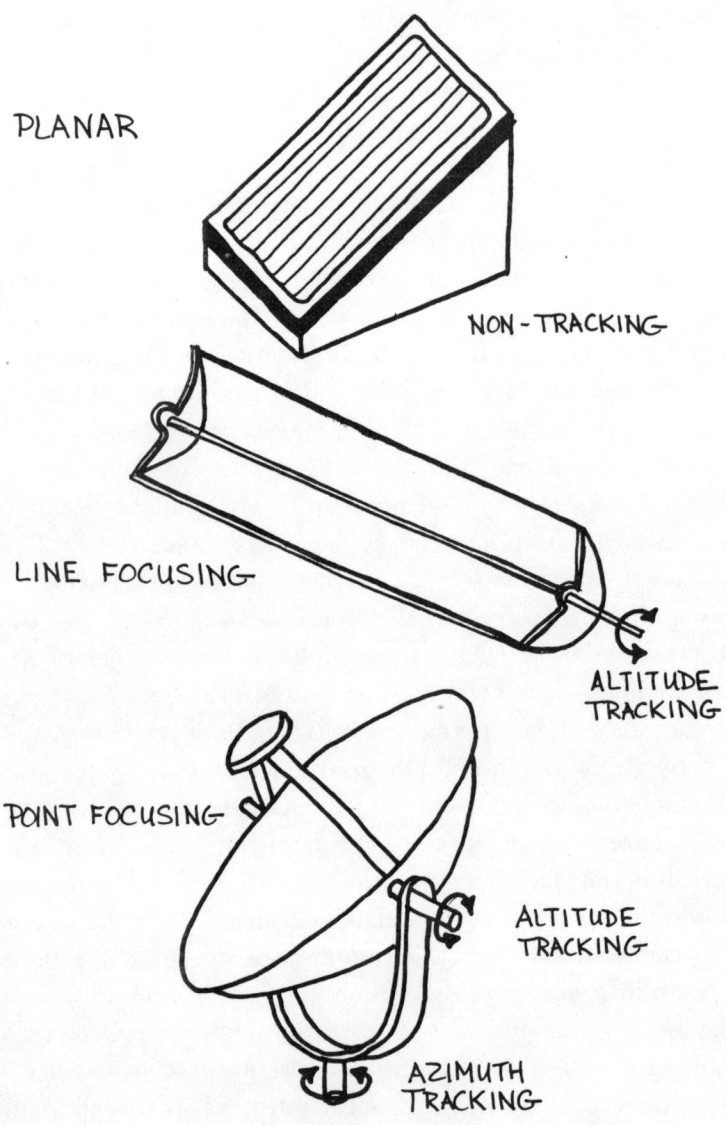

FIGURE 6

2: THE ALTERNATIVES

The quality and quantity of solar radiation varies widely from one location to another. The techniques used to capture this radiation also vary. We have already briefly examined one method: passive solar collection. This means of collecting and using solar energy will work in any locality regardless of radiation quality or quantity. It is basically a process which transfers low grade heat according to the laws of natural convection, conduction and radiation. This low grade heat is typically in the range of 60 to 100°F. If we upgrade our heating requirements to a range of 100 to 200°F then the operating parameters of the solar collection system become more stringent. Moving the energy becomes a problem unless fans and pumps are included. Controlling the energy, to be released only when needed, requires the addition of electronic regulating devices. As the systems grow more complex, the versatility of the energy derived from them increases.

When a solar system employs fans, pumps or mechanical devices to operate it, the system is termed "active." The most important component of an active solar system is the collector. These units come in many sizes, shapes and forms. Each different type has been developed to meet a particular requirement, whether it be 140°F water for domestic purposes or 1200°F water for steam generators.

COLLECTOR TYPES

Solar collectors may be placed into three general categories:
1. Point focusing collectors
2. Linear focusing collectors
3. Planar collectors

These three categories are illustrated in Figure 6, Solar Collector Types.

The *point focusing collector* usually consists of a spherical mirror which reflects and concentrates all the sunlight which hits its surface onto an absorber located at the point of focus. If a fluid is then circulated through the absorber it will contain and transport the heat away from the collector area to a storage area. Because all the energy incident on the collection surface is focused on one central point, the operating temperature is quite high. A spherical collector 30 feet in diameter may generate super-heated steam in the area of 300 to 400°F at only 12 lbs. per square inch pressure.

A similar system consisting of hundreds of flat mirrors all set to reflect on one central tower may provide high pressure steam to power generators. This device, termed the "power tower," is currently being explored in southwestern regions of the U.S. and has some potential for electric generation. Large scale power plants of this nature, however, are in the infancy stage and are far from being commercially practical.

In general, point-focusing collector systems are not practical for residential use. Sophisticated tracking systems are required to match solar azimuth and inclination. These tracking systems are both expensive and difficult to find on the commercial market.

Linear focusing collectors are similar to point focusing systems in that a reflective surface is employed to concentrate the radiation on to an absorber. In this case, however, the area of focus is a line instead of a point. The absorber consists of a long tube located at the focal point. Again, fluid moving through the absorber tube is heated and transferred to storage. Temperatures above 200°F are easily attainable. This allows the linear focusing collector to be especially attractive for air-conditioning applications where

2: The Alternatives

high temperature fluid is required to drive absorption chiller units. Tracking systems are still required, though the degree of sophistication has been lowered somewhat because only the solar inclination need now be matched. Change in azimuth will have little effect as the collector has a linear component.

The overall drawback to both point and linear focusing collectors is that direct radiation must be available to drive the units. Scattered radiation will strike the collector at all angles and

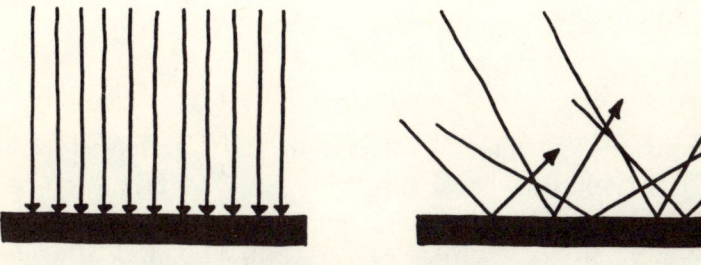

FIGURE 7

disrupt the re-focusing process. Figure 7 shows the difference between direct (columnar) and diffuse (scattered) radiation and their effect on reflective surfaces. This basic principle is important in the New England region because a large percentage of the radiation is diffuse in nature. In southern areas such as Florida, Texas and Arizona the direct component is more common and focusing collectors are quite practical.

Figure 8 indicates the relationship of diffuse to total energy. The horizontal scale is the ratio of total daily radiation to total extraterrestrial radiation. In simple terms, this is the fraction of the sun's energy which strikes the earth directly, compared to the fraction which is scattered due to cloud cover, particulate matter and atmospheric inconsistencies. For instance, the graph can be used to determine the amount of daily radiation that is

Solar Energy for the Northeast

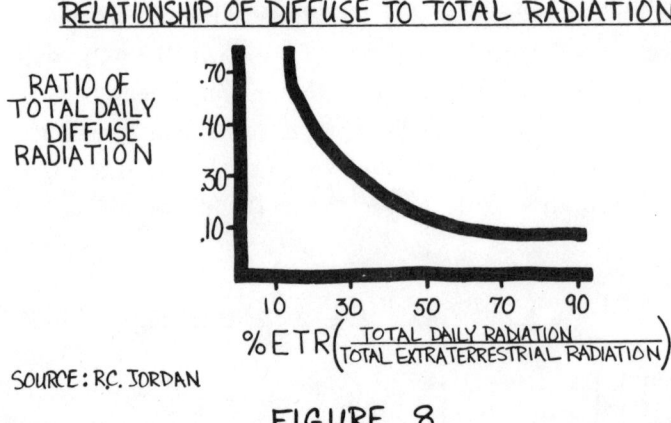

FIGURE 8

scattered if the % ETR is 30 and the total radiation (from figure 5) is 1,400 Btus per square foot. A 30% ETR corresponds to a ratio of .50 diffuse to total energy. 1,400 × .50 = 700 Btus per square ft. diffuse for the day. The direct radiation is then (1,400 − 700) 700 Btus per square ft. per day.

It is not necessary that you do these calculations before installing your own system. The graph instead is intended to emphasize the fact that the diffuse component is significant; this significance becomes further underscored by the fact that in the Northeast, the radiation is substantially diffuse.

The overall conclusion is that in areas of high diffuse energy, focusing collectors in either the point or linear category, are not advisable. Their high temperature capabilities are certainly desirable but not attainable on a consistent basis in the Northeast United States.

The remaining, and most logical, alternative lies in the third category: *planar or non-focusing solar collectors*. A planar collector, otherwise known as a flat-plate collector, employs no reflective surfaces which are dependent on direct radiation. The entire collector surface is the absorber and all or part of this surface has a fluid which moves across it to contain and transfer the energy.

2: The Alternatives

The flat-plate collector is by far the most widely used and commercially available type of solar collection device. No tracking systems are required and the cost is relatively low when compared to focusing type systems. In the Northeast, the flat-plate collector is the only practical means of capturing energy in active solar systems.

Planar collectors are now available from well over 200 manufacturers in this country. The quality and efficiency of these manufactured units vary considerably, and it is unfortunate that the average homeowner, when shopping for solar collectors, places his highest priority on collector cost. There are many areas where a collector manufacturer can cut corners to provide an inexpensive unit, but whenever a corner is cut, the quality of the unit decreases.

If a solar collector is rated in dollars per square foot, then no attention is given to performance in terms of energy output. If a collector is rated in dollars per Btu, it then reflects how much energy you can expect to receive given a specific investment. This is not practical from the manufacturer's perspective because the output of a given solar collector will vary widely between locations, dependent upon the solar availability. The manufacturer can not possibly provide this data without years of research into climatic conditions of all the areas in which the collectors will be installed.

Given this situation there remain two criteria upon which you can judge solar collectors. These will remain regardless of the location or the application of the unit.

1. Sustained efficiency
2. Structural integrity

The efficiency of a flat-plate collector is a result of the design and the materials used in its construction. We will look at this in great detail in Chapter 3, Solar Collection. How the collector will perform in year 20 compared to year 1 is the present point of interest. Many units having high initial operating efficiencies soon lose their ability to capture and transfer energy, or at least that ability becomes seriously impaired by poor design and cheap materials. These collectors require annual maintenance in order to maintain effective operation. Such items as glazings, insulation, and seals may deteriorate rapidly when exposed to a constant heating and cooling cycle 365 days a year. If you are preparing to make a substantial investment in solar heating equipment you will certainly want components that will stand up for 20 years of operation and sustain their high efficiency.

The structural integrity of a collector is closely associated with efficiency. If the collector is to last it must be durable. Temperatures ranging from 0-400°F may occur in a flat-plate collector, often within a few hours, during winter operation. This thermal shock will affect all materials involved in construction. Metals, plastics and glass will expand and contract; insulation, gaskets, and coatings may slowly oxidize and give off a gas; heat transfer fluids may boil and lead to acid formation and scaling. All of these problems have a solution and you must determine if the manufacturer is aware of them.

If you are building your own collectors, steps can be taken to minimize these obstacles, though you may find that totally eliminating them is not practical. In some cases, the materials are not readily available to homeowners, but are accessible to manufacturers who can purchase in production quantities. For instance, low iron tempered glass can be difficult to buy in lots of less than 5,000 square feet. Unless you're planning on building a very large greenhouse you will probably not want to take this route. An inexpensive but readily available plastic glazing will serve the same purpose provided you are willing to replace it periodically.

In any event, there is no reason why a manufacturer cannot use the best available materials. The collectors, of course, will cost more than a homemade version, but they will also last longer. The manufacturer is now required by the Federal Department of Housing and Urban Development to guarantee solar collectors for a minimum period of five years. This gives the homeowner an opportunity to observe the system and note any deterioration in its effectiveness. There have been collectors commercially sold which have lost up to 50% of their effectiveness inside of two years. Generally, these manufacturers do not remain in business

2: The Alternatives

long and the guarantee period is meaningless. On the other hand, there are also many manufacturers whose collectors have been installed and operating consistently for better than five years.

Unfortunately, the energy crisis is relatively new and this has created a situation where few commercial solar collectors have a twenty-year history of operation to fall back on. In the 1920s flat-plate collectors were being produced in Florida but these were somewhat crude as compared to the current state-of-the-art solar collector.

The industrialized art of manufacturing solar collectors did not re-emerge until early in the '70s. During the first half of this decade, the groundwork was laid, prototypes were built, and test installations monitored. Serious manufacturers overcame the obstacles and re-designed to meet higher standards. The result was the emergence of efficient and durable flat-plate solar collectors for the residential market.

APPLICATIONS

A solar collector makes hot water; this fact is inescapable. During a demonstration in upstate New York in May of 1974, I first witnessed the ease with which even a homemade collector could produce hot water. Two crude solar collectors were connected by a rubber garden hose, fastened with stainless steel hose clamps. At 11:30 a.m. the temperature of the water being used to cool the collector plate had reached 185°F. In another hour, the garden hose burst from the high pressure and delivered billowing columns of steam to the shocked spectators. The point had been made; solar hot water was no technological mystery.

There are many uses for solar hot water and hot air. Which use is most practical at your site will depend on your requirements. The following is a brief summary of the alternatives available.

Domestic Hot Water: The private home is a notorious consumer of hot water for bathing, dishwashing and laundering; demands of up to 20 gallons per person per day are considered average. Solar systems can provide up to 80% of this requirement on an annual basis. Storage tanks of sufficient quantity can maintain a substantial portion of the hot water for use on cloudy days. It is surprising to realize that you may require between 300 and 800 kilowatt hours of electricity to do the same job on a monthly basis. This translates to between $15.00 and $40.00 at the current average rate of 5¢ per kwh. With rising costs of electricity the potential savings increase dramatically. Solar domestic hot water is unmistakably the most cost-effective alternative.

Space Heating: In general this is the largest consumer of energy in the home. If you have an existing home, retrofitting it with a solar collection system can be difficult and expensive. Collector area and storage area must be large. For these reasons, solar heating is most practical in new construction. The use of passive collection must be designed into the home by the architect or builder. Active systems including flat-plate collectors must be incorporated into the building facade to insure an aesthetic final design. After all, not only do you have to heat the home, but you will live in it as well. Large expanses of solar collectors can be only as attractive as you make them.

Systems can be sized to provide between 50% and 80% of the annual heating requirement. This is accomplished by either water or air heating solar equipment; in either case a conventional forced air back-up system will be required. By all means, do not assume that because a collector produces hot water, a baseboard hot water system is the proper choice. It is not. A typical baseboard system produces heat by radiation from 180°F water in a closed recirculating loop. The temperature in this loop seldom drops below 140°F. The typical solar system is producing at 140°F or lower, and provides little help to a baseboard system.

In a forced air system, water can be circulated through a coil with a fan blowing across it to remove the heat and transfer it to the living space. This is heating by convection and requires medium temperatures of 100° to 120°F. Another alternative is heating by conduction through the use of coils in the floor. This low temperature system works effectively with 80 to 100°F operating temperatures and is quite satisfactory for people who are happy as long as their feet are warm. A closer look at these alternatives will be presented in Chapter 6, Systems.

The overall decision to use solar for space heating will probably rest on the cost of the equipment and installation. Home-fashioned systems may cost 6 to 12 thousand dollars and require maintenance while manufactured systems are available in the 12 to 20 thousand dollar range. Indeed, this is a high initial cost, but consider also that it requires almost $2,000.00 per year to heat your home by conventional oil and electric sources at current rates.

Solar Greenhouses: The greenhouse acts quite effectively as a passive solar collector. Heat trapped in this enclosure can be stored in 55-gallon drums of water and later passed by convection into the home through the use of a fan. Or the fan can be run all day long to eliminate the storage step and transfer the heat directly into the living space during the day. In either case, the greenhouse itself will certainly remain warm enough to maintain garden crops during the winter and early-starting plants in the spring.

Solar Dryers: The crops you raise in the greenhouse during the winter or in the garden during the summer can be quickly dehydrated through the use of a simple solar dryer. These vegetables, when reconstituted with water, are often more nutrient than they would have been had conventional canning or freezing methods been used. An inexpensive, easily built food dryer will reach temperatures as high as 200°F which is sufficient, enough to dehydrate a plump juicy tomato in only 12-20 hours. Admittedly, this is considerably longer than the usual canning process, but overwhelmingly less labor is involved.

Solar Swimming Pool Heating: A comfortable temperature range for swimming pools is 70° to 90°F. If you recall that solar collectors are most efficient at low temperatures, then this application of solar energy becomes readily apparent. Add to this the fact that your requirements will occur mostly during the spring and fall when freezing in the collector is not a problem. This means that you can circulate water directly from the pool to the collectors. It is not at all difficult to overheat a pool during the summer by leaving the solar system on. Care should be taken to include a bypass loop unless you intend to soft-boil 2,000 eggs in your swimming pool. If the pool is enclosed and a substantial pool blanket/cover system is installed, it is possible to maintain 70°F water all year round for swimming pleasure. In this case, a heat exchanger will be necessary as you can no longer circulate water through the collectors. Low cost manufactured systems can be quite dependable for this application as well as inexpensive home built units.

The uses for solar energy are by and large up to your own creativity. The five alternatives listed above are the most common applications. These are tried and true uses which have been employed in the Northeast successfully. Sample system designs for each of the alternatives are included in Chapter 6.

In choosing between these alternatives, you must first carefully determine what your requirements will be. A house which includes all of the above options will truly become a solar home and no one can debate you on that point. However, it is most likely that financial resources will limit your usage to one or two of the above alternatives. In that case, the priority item becomes domestic hot water. It is affordable to most people and will displace the most conventional fuel in proportion to the investment. In many cases, the best choice is to install a 3 or 4 collector hot water system to familiarize yourself with solar equipment. Based upon this evaluation, you can then assure yourself that a space heating system is also cost-effective. In retrofit situations, six or eight collectors may be used to heat a living or family room only. Daytime solar combined with night time wood burning takes care of this requirement nicely.

The solar greenhouse, drying and pool heating options can always be added on to the home after construction, if you plan accordingly. These are typically smaller investments with low

2: The Alternatives

installation costs. Overall costs on the hot water and space heating systems can be reduced substantially by Federal aid to homeowners on energy conservation measures. Many additional alternatives are available to help defray the cost of solar equipment. These vary from state to state and may require that you contact your state energy office to determine which are applicable. You should be aware of the possibilities, however, and therefore the following section is included on options for financing your system.

FINANCING SOLAR HOME IMPROVEMENTS

The purchase of solar energy equipment is an investment in energy security. It is an investment which is substantial as well as wise. At the writing of this book the cost of #2 fuel oil had reached $1.00 per gallon. There is no reason to believe that it will not continue to rise dramatically. Conservative estimates indicate that at least a 12% per year escalation rate may be expected. Although this does not appear drastic at first, a look at a 10 year

TEN YEAR FUEL COST PROJECTION

YEAR	COST OF FUEL OIL
1981	$1.12/GAL.
1982	$1.25
1983	$1.40
1984	$1.57
1985	$1.76
1986	$1.97
1987	$2.21
1988	$2.47
1989	$2.77
1990	$3.10

FIGURE 9

Solar Energy for the Northeast

projection is shocking. Figure 9 indicates the frightening results.

By 1990, fuel oil costs may well reach over $3.00 per gallon, and the cost of heating an average size home (2,000 square feet) will be greater than $3,000.00 a season. At this level, a solar system which costs $15,000.00 will recover its investment in 5-7 years. Of course, by 1990, systems will probably cost more than this, which is a good reason to start planning your solar home now. If the present cost of fuel prevails, a system may require 15 to 18 years to demonstrate its cost-effectiveness. The question is, just how confident are you that the present cost will prevail; not very! Considering that the major portion of our oil is imported and that the price of oil is being deregulated the outlook is indeed bleak.

The only choice is to use less oil.

The installation of active solar collection systems requires a high initial investment, ranging from 5 to 20 thousand dollars. Generally this home improvement must be financed through a bank, credit union, or similar lending institution. The financing of solar improvements is a rapidly changing situation. State legislatures, in many cases, have either enacted or are considering legislation to aid in defraying these costs to the homeowner. Instead, there are several general areas which you may look into. Your state energy office should have up-to-date information on the status of these programs in your area. These offices are listed in Figure 10.

Solar Bank Loans—A solar bank is an entity enacted by state or federal legislation to make accessible long term, low interest loans to individuals installing energy conserving devices. The state may issue bonds to raise the capital and then redistribute it at 6-8% interest for periods up to 10 years.

At the state level several bills to create solar bank loans are being considered. Connecticut enacted a solar bank in August of 1979 and many New England neighbors are certain to follow. Check with your local energy office.

The most attractive point of a solar bank loan is that you will often find the monthly cost to service the loan is less than the

2: The Alternatives

FEDERAL AND STATE ENERGY OFFICES

NATIONAL SOLAR HEATING AND COOLING INFORMATION CENTER POST OFFICE BOX 1607 ROCKVILLE, MARYLAND 20850	
NORTHEAST SOLAR ENERGY CENTER 70 MEMORIAL DR. CAMBRIDGE, MASSACHUSETTS 02142	
CONNECTICUT O.P.M. ENERGY DIV. 20 GRAND ST. HARTFORD, CONNECTICUT 06115 203-566-3395	NEW YORK ENERGY OFFICE EMPIRE STATE PLAZA-AGENCY BLDG 2 ALBANY, NEW YORK 12223 518-474-1785
MAINE OFFICE OF ENERGY RESOURCES 55 CAPITOL ST. AUGUSTA, MAINE 04330 207-289-2195	RHODE ISLAND ENERGY OFFICE 80 DEAN ST. PROVIDENCE, RHODE ISLAND 401-277-3370
MASSACHUSETTS SOLAR ACTION OFFICE RM. 1413 - ONE ASHBURTON PL. BOSTON, MASSACHUSETTS 02108 617-727-7297	NEW HAMPSHIRE ENERGY COUNCIL 26 PLEASANT ST. CONCORD, NEW HAMPSHIRE 03301 800-562-1115
VERMONT STATE ENERGY OFFICE MONTPELIER, VERMONT 05602 802-828-2393	

FIGURE 10

amount you would have spent on conventional energy sources. For instance, on a $2,000.00 loan, at 8% interest, the monthly payments are $24.28 over a 10 year period. If you are spending $25.00 a month to amortize the loan and at the same time saving $30.00 a month due to reduced fuel consumption you have actually cut your expenses at the very onset. There is then no need to wait for your energy savings to cover the cost of the investment and the notion of 'payback period' is no longer applicable.

State Sales Tax and Property Value Assessment—Sales tax on the purchase of alternative energy equipment is quickly being phased out. Not only the solar collectors, but the ancillary equipment and even the installation cost may be tax exempt. Your state tax office will have current information on this.

The market value of your home will increase if you add on solar energy features. The fuel consumption of a structure has become a prime consideration in home purchases. You can consult your local real estate broker on this point. Although the market value increases the assessed value may not. On most home improvements, the town assessor will determine the value of the improvement so that you are taxed accordingly. Again, solar renovations, in many cases, are exempt.

The National Energy Act—In the fall of 1978, after years of concentrated effort, the United States developed its first comprehensive energy plan. Most of the act is concerned with the use of our major conventional energy resources: oil, coal and natural gas. However, included in this compilation of deregulation structures and administrative services, are the residential energy conservation tax credits. These tax credits are quite significant and by far the most encouraging legislation yet enacted in favor of solar energy alternatives.

The credit is available to anyone who installs solar equipment and pays an income tax. A full 30% credit is available for the first $2,000 you spend and an additional 20% credit for any amount over $2,000. A ceiling of $2,200 in tax credits is allowable to an individual in any one year. For example, if you are installing a solar hot water system which costs $3,000, you are eligible for 30% of $2,000 or $600 and 20% of the additional $1,000 or $200. The net savings is $800 which reduces the cost of the system from $3,000 to $2,200. That's a saving of over 26%.

In general, the financing picture for solar investments is improving rapidly. Many banks have individual programs apart from the standard home improvement loans. Solar is no longer being considered as a risk, but rather, a hedge against inflation. The lending rate you obtain must be carefully evaluated to insure that it will remain within your capabilities. Spending the extra time to locate a low interest solar bank loan is certainly worth while; money available at 6% interest is exactly twice as attractive as money at 12%.

3: SOLAR COLLECTION

The science of solar collection is subject to several physical retraints. The most obvious of these is the quality and quantity of the radiation transmitted by the sun. The effect of the atmosphere and the terrestrial environment will also influence the manner in which this radiation finally strikes the solar collector. Many design factors must be taken into consideration in order to maximize the potential energy gain and minimize losses in every phase of the collection process.

COLLECTOR AZIMUTH AND INCLINATION

These two variables can make a world of difference in the performance of a solar collection system. It makes sense to assume that if you are trying to capture sunlight, the collector must be oriented in such a manner as to optimize the collected energy. This is an area of system design which is often characterized by "general rules," the two most predominant of which are:
    1. Face the collectors due south.
    2. Adjust the inclination to the latitude plus 10°.

Solar Energy for the Northeast

In actuality, the orientation of the collectors will depend on the application for which you are using them and the time of year during which you will expect the best performance. The sun's position in the sky is constantly changing in small increments all year long. As we have discussed, it is not practical to design a mounting frame which will allow you to track the path of the sun; the cost of these devices is high as well as the maintenance of the drive mechanism. Because the flat-plate collector is effective with diffuse radiation, it is not necessary that the collector plane be always perpendicular to the incoming rays. Instead, you will wish to choose a fixed position which provides the best all year round average performance.

In determining the position to face the collector (azimuth), it is readily apparent that south offers the best results. In most New England areas, the sun is more intense in the afternoon than in the morning. This is because it may take several hours of sunlight to heat the atmosphere and clear away morning mist, fog and haze. In that case, it is advantageous to shift the collector azimuth slightly west of south to capitalize on more intense afternoon radiation. This shift should only amount to about 10° for the best results. If you swing greater than 20° west of south, collector

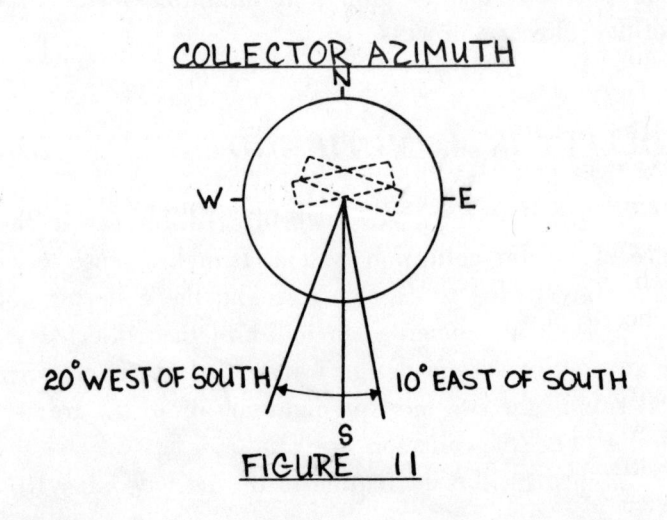

FIGURE 11

3: Solar Collection

performance will begin to drop off quickly as the morning sun, rising in the east, will be almost totally lost.

Figure 11 illustrates the western adjustment necessary to overcome poor transmission of morning sun. When determining the direction of due south remember that it is true south and not magnetic south which you are considering. In New England, these deviations range from 8° West in Binghamton, New York to 20°

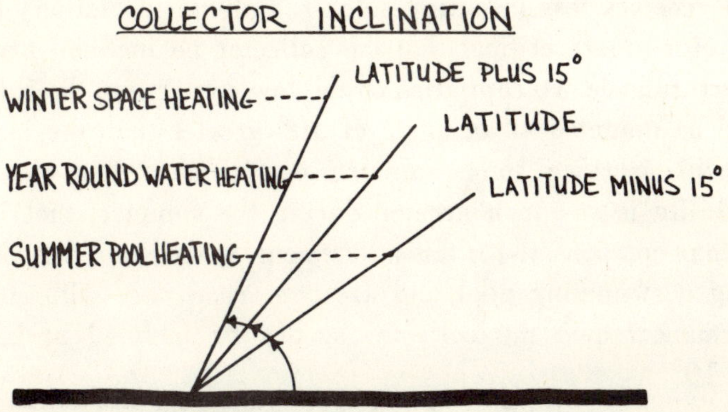

COLLECTOR INCLINATION

WINTER SPACE HEATING — LATITUDE PLUS 15°
LATITUDE
YEAR ROUND WATER HEATING
LATITUDE MINUS 15°
SUMMER POOL HEATING

FIGURE 13

West in Bangor, Maine. Figure 12 summarizes these variations for other areas. Once you have determined how much the deviation from true south is at your site, you must also consider the effect of local deviation. A large metal source near at hand may throw your compass off an additional 3-5°. It is also advisable to remain at least 50 yards from automobiles or building structures while observing a compass. The quality of the compass which you are reading is also a determining factor.

Once you have located south on the compass, corrected for the deviations listed above, and removed yourself from the influence of surrounding metal objects, you can then adjust an additional 10° west of south and locate solar collectors with confidence.

Following the previous exercise, the inclination or tilt from horizontal must be determined; this is an attempt to locate the

Solar Energy for the Northeast

plane of the collector in a position perpendicular (normal) to the sun's rays. Because the sun's altitude changes, it is not possible to maintain this angle of solar incidence constantly. Again, the best possible average position is the goal.

Latitude will aid in the determination of collector inclination. Because nearly ⅔ of the total annual energy occurs in the summer, you can be fairly confident that collectors will perform well during this season when the sun is higher in the sky. The winter receives less incoming solar radiation (insolation) and it is therefore more critical that the collector be inclined properly for performance. To capitalize on the low winter sun, the collector should be mounted at an angle of 10° greater than the latitude. This will sacrifice some summer performance, but the sun's availability is so much greater during the summer, that it will more than compensate for this adjustment. In the event that you are heating a swimming pool and are concerned only with summer performance, then the collector should be inclined at latitude minus 10°.

MAGNETIC DEVIATIONS IN THE NORTHEAST

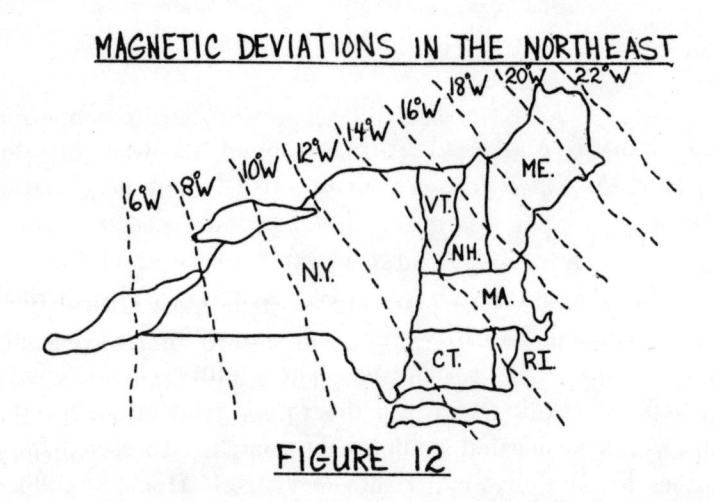

FIGURE 12

Figure 13 indicates the correct angle for collector performance as a function of application. If it is possible to remain within 5° of these recommendations, you can be assured that the performance will remain optimum with only minor deviations.

Having the solar collector correctly oriented to azimuth and inclination insures the best possible starting ground. The solar radiation falling on the collector plane is now at its maximum and the performance of the system, including its efficiency, will be a function of the collector design and system configuration.

COLLECTOR DESIGN

A solar collector is a device having two primary objectives:
1. Maximizing heat gain
2. Minimizing heat loss

If the heat losses could be entirely controlled so that no energy entering the collector was reflected or re-radiated, then the collector would be 100% efficient. Many technological innovations have been explored to pursue this goal. Currently, the maximum attainable collector efficiency is around 80%, which is a considerable accomplishment. In all likelihood, the average operating efficiency of a manufactured collector will be in the range of 60 to 70%, while a home built version may reach a 50 to 60% efficiency.

The variables involved in designing a solar collector to operate at peak efficiency are complex and interrelated. In order to make a thorough examination of these elements, we will look at the basic components of a flat-plate solar collector and the various design alternatives.

The Collector Plate—The plate is the single most critical item in the solar collector. Its purpose is to absorb shortwave solar radiation, convert it to long wave radiation and transfer it to a fluid media. The myriad of plate configurations attest to the research devoted to perfecting efficient, high temperature collector plates.

Figure 14 shows several common examples of plate designs. By far, the most widespread design is the fin and tube type which incorporates a flat fin for the absorption of radiation. When the fin gets hot, the thermal energy moves by conductance to the lower

Solar Energy for the Northeast

HYDRONIC PLATE DESIGNS

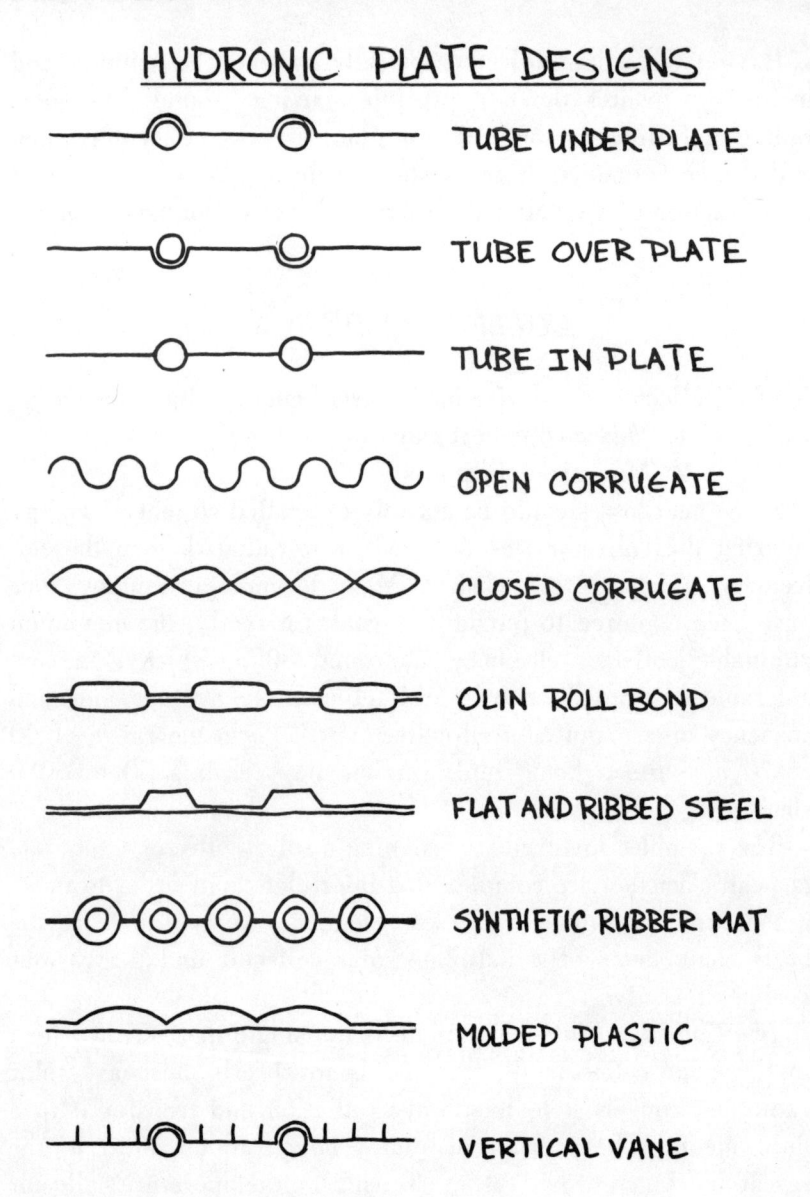

- TUBE UNDER PLATE
- TUBE OVER PLATE
- TUBE IN PLATE
- OPEN CORRUGATE
- CLOSED CORRUGATE
- OLIN ROLL BOND
- FLAT AND RIBBED STEEL
- SYNTHETIC RUBBER MAT
- MOLDED PLASTIC
- VERTICAL VANE

FIGURE 14

3: Solar Collection

THERMAL CONDUCTIVITY OF SEVERAL METALS

	TEMPERATURE °F	CONDUCTIVITY (k)
ALUMINUM	212	119.0
COPPER	212	218.0
STEEL	212	25.9
NICKEL	212	34.0
TIN	212	34.0
BRASS	212	68.3
IRON	212	36.6

FIGURE 15

temperature tube where a fluid is used as the transferring absorptive agent. In an air type collector, the tubes are eliminated and air moving across the surface of the plate absorbs and transfers the energy.

In an effort to enhance the absorptive nature of the collector plate, a suitable metal with qualities that encourage conductance is necessary. Copper is the best alternative available for this purpose. The conductance of a metal is given by the coefficient "k" and conversely its resistance to heat is given by the coefficient "R" which is a function of $1/k$. Figure 15 lists the conductance and resistance of metals currently used in collector plates.

The cost of these metals plays an important role in determining collector design. Although copper produces the best results, it is also among the most expensive. An entirely copper collector plate represents the greatest possible values for thermal absorption and conductance. The thickness of metal will also influence its ability to conduct heat. As illustrated in Figure 16, a copper plate of .020 inches thickness will be less efficient than an aluminum plate of .040 inches. In contrast, if the copper is .040 inches also, its efficiency is about 4% higher. These values can be offset by including more tubes on the plate surface and thereby reducing

Solar Energy for the Northeast

VARIATIONS IN ABSORBER PLATE EFFICIENCY

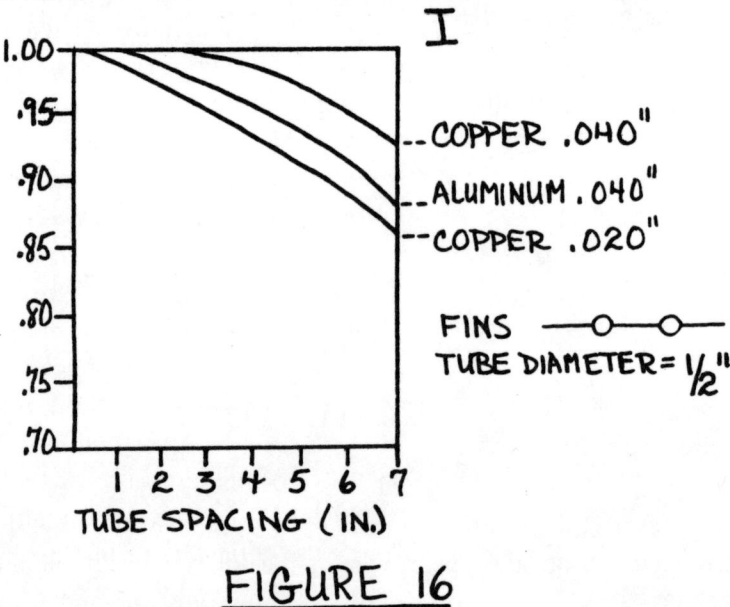

[Graph: Y-axis from .70 to 1.00 in increments of .05; X-axis "TUBE SPACING (IN.)" from 1 to 7. Three curves labeled: COPPER .040", ALUMINUM .040", COPPER .020". FINS ─o──o─ TUBE DIAMETER = 1/2"]

FIGURE 16

the distance that heat will have to travel to reach the cooling fluid. Figure 17 also demonstrates that a steel plate with a thickness of .040 and tubes ½ inch in diameter spaced 5 inches on center will be slightly more efficient than an aluminum plate of .020 inches thickness with ½ inch tubes spaced 7 inches on center.

The number of tubes on a plate will also have an influence on the amount of fluid moving through the collector and the percent wetted surface of the plate. If an entirely 100% wetted surface is employed, then radiation striking the plate surface will be transmitted directly to the fluid thereby minimizing re-radiation losses which occur in the fin during the conductance process.

In an air type collector where no liquid passages are necessary, the entire collector plate is essentially the wetted surface. Several hydronic type collectors employ a waffle designed plate in which greater than 95% of the surface is wetted. The waffle design characteristically requires more metal in its construction and therefore represents a costly alternative if you wish to fabricate

3: Solar Collection

the plate out of copper. Generally, waffle type construction employs aluminum or stainless steel as the conducting material.

The corresponding improvement to maximize wetted surface in an air type collector is to include vertical vanes. By increasing the surface area, the air will be exposed to more potential heat gain. Care must be taken in a vertical vane air collector to insure that the vanes do not cause undue shading of each other or the horizontal portion of the collector plate. Due to this shading effect a vane type air collector is only about 4% more effective than a flat type collector. Its construction is illustrated in Figure 18.

The collector plate should also encourage turbulent flow of the cooling medium. In a hydronic unit there is not much you can do to enhance turbulence other than choose a fluid which is characterized by turbulent flow. A greater discussion on this will be found in the section on heat transfer fluids. The air collector, however, can employ several baffles to trip the air flow and cause mixing as the air moves through the collector. In this manner the

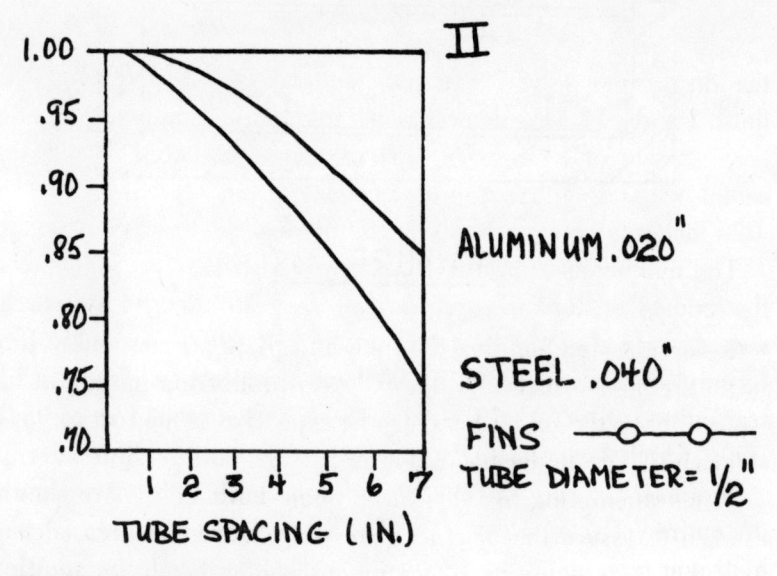

II

ALUMINUM .020"

STEEL .040"

FINS ─o──o─
TUBE DIAMETER = 1/2"

TUBE SPACING (IN.)

SOURCE: R.W. BLISS

FIGURE 17

Solar Energy for the Northeast

air is heated more evenly and its capacity to absorb thermal energy increases.

The tubes or channels which carry the cooling fluid may be located in several different configurations. In the classical hydronic collector, the tube is generally ⅜ to ½ inch in diameter and is bonded by a solder joint or press fitted directly to the plate. It is very important that the bond between the tube and the plate have a very high coefficient of thermal transfer. A poor bond, such as a spot-soldered tube, will cause the fin to heat up without transferring the energy to the tube. Typically, home built systems employing a pipe retainer to fix the tube to the plate result in low operating efficiencies.

VANE TYPE AIR COLLECTOR

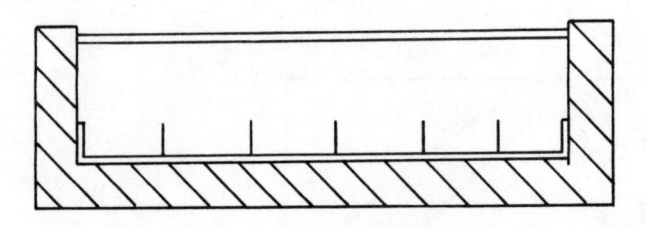

FIGURE 18

The options for bonding the tube and plate are extremely limited to the homeowner who doesn't have production tooling at his disposal. Even the solder bond can be expensive as the cost of 95/5 solder rises. The solder bond, however, is the best possible alternative when attempting to fabricate a home built unit. Care should be taken to assure that surfaces are well cleaned, degreased and fluxed before attempting a continuous solder bond. In addition the heat of the torch will have a tendency to cause thin metal fin material to warp and buckle as the solder is applied. For this

3: Solar Collection

reason, heavier gauge metals must be used to insure adequate and continuous bonding.

In many cases, the collector tube material may differ from the fin material. Typically the tube is copper while the fin is aluminum or galvanized metal. When bonding dissimilar metals, the problems of galvanized corrosion must be thoroughly considered. These will have an effect on the maintenance and replacement schedule of home built units. Manufactured collector plates, by and large, avoid this problem by using the same metal throughout the plate.

Locating the fluid channels is a significant design question. Tubes fixed to the top of the plate will absorb more energy because the tube itself is part of the absorbing surface. In contrast, tubes located beneath the plate will rely totally on the heat of conductance to acquire their thermal energy. Tube-in-sheet type plates offer the superior alternative, though, again, these are not likely to be available to the do-it-yourselfer. The tube-in-sheet process has been most successfully accomplished by the Olin Brass Corporation in their patented Roll-Bond collector plate. In this case, two metal plates are silk screened with a channel pattern and then fused together under high temperature and pressure. The screened portions do not fuse and are later expanded to form a fluid channel within the plate itself. The result is a collector plate without severe heat transfer and corrosion problems.

Another design employs open fluid channels where the liquid is exposed to the air. This configuration generally employs a sheet of corrugated metal as the plate, with water trickling down the corrugations. Special attention must be given to the flow rate of the fluid in order to insure that evaporation off the plate and condensation on the back of the glazing does not occur. If condensation is allowed to collect on the glazing, it will seriously inhibit the transmittance of solar radiation through the glazing material. The flow rate, on the other hand, cannot be too fast or the fluid will not be sufficiently heated. Attempting to strike a workable medium can take considerable time and testing.

Solar Energy for the Northeast

HYDRONIC FLOW CONFIGURATIONS

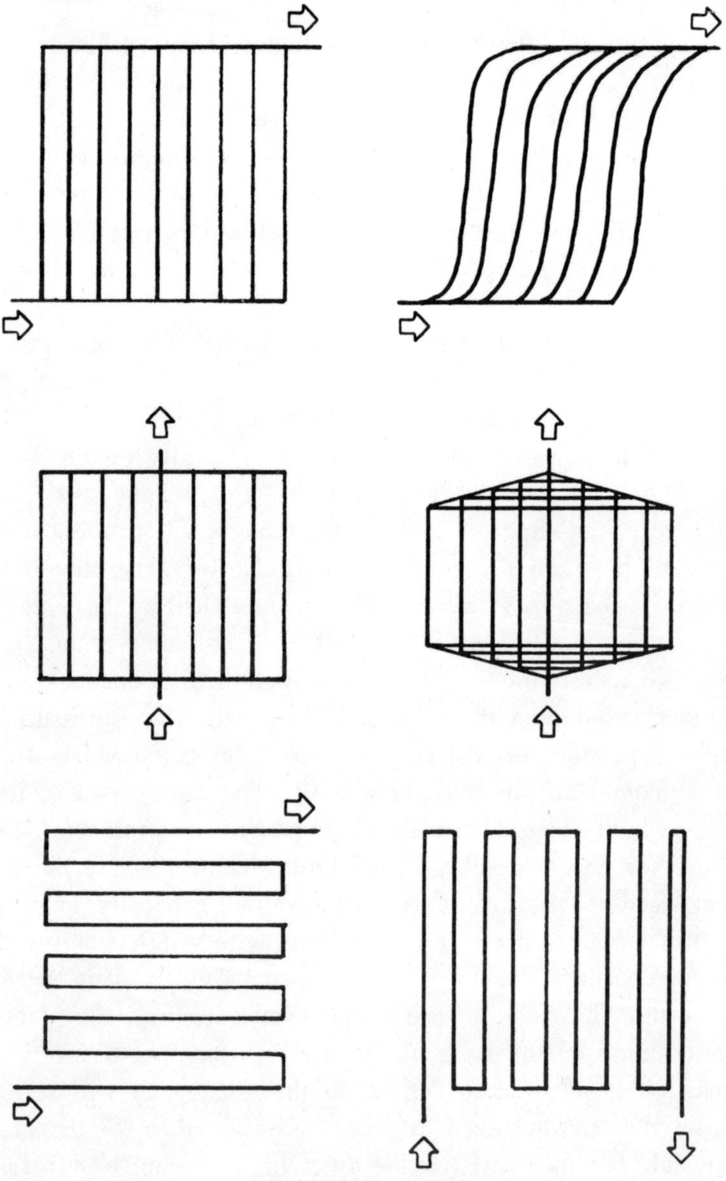

FIGURE 19

3: Solar Collection

The configuration of the flow pattern will influence the temperature which a collector plate can attain. The desired effect is to keep the plate as cool as possible. In this manner, fluid entering the collector absorbs the maximum amount of heat and transports it quickly away to storage. Various tube configurations are shown in Figure 18. Design considerations should include uniform fluid flow, low pressure drops and easy construction. The plate will generally consist of several vertical tubes called risers which are connected at bottom and top by a horizontal tube called a header. The risers are typically between ⅜ to 1 inch in diameter. In order to insure uniform flow rates through all the risers, the plate must be properly manifolded. The classical configuration is by far the best means to overcome uneven flow. Sinusoidal tube patterns are capable of attaining higher temperatures, though the overall plate efficiency is seriously limited. A collector plate operates most efficiently at low inlet and outlet temperatures. A closer examination of this will be given in the section on collector efficiency curves. In the sinusoidal design the fluid moves a greater distance across the plate and is therefore working at a higher temperature before it has passed halfway through the collector. During the passage of the remaining half, the heat that is absorbed will be substantially less.

When connecting several plates together in a sytem, a parallel pattern is the rule. Arranging collectors in series will produce the same problem with overheating as described in the sinusoidal tube configuration. Figure 19 indicates the parallel system configuration.

The collector plate temperature may range from below zero on cold winter nights to 450°F during the day under no-flow conditions. This broad temperature range brings us the design problem of overcoming expansion and contraction. Figure 20 indicates the expansion and contraction coefficients of several metals used in plate contruction. A plate without means to expand and contract during its lifetime will inevitably have a shorter life. This minor movement repeated daily over a 10-year period can lead to small cracks and distressed areas in which leaks originate. The problem is not difficult to overcome. Many fittings are available to provide for movement in the plate. These fittings are typically a bellows or coiled fluid passage located at the collector inlet or outlet. The plate is maintained in a "floating" position by the expansion fitting.

When mounting the plate in the frame, it will be necessary to attach it to at least one point to prevent it from shifting and moving around inside the enclosure. This attachment should provide a method of thermally isolating the plate from the frame. If the metal plate comes in direct contact with the frame, then conductive heat losses will occur at this point. Although a minor problem, thermal isolation is easily accomplished through the use of an insulating bushing. A ceramic, phenolic or similar material may be employed. Figure 21 indicates a typical method of thermally isolating the plate from the frame.

Many factors must be thoroughly considered before designing an efficient solar collector plate. The following check list provides a review of the most important aspects as discussed in this section.

Plate material	Tube spacing
Plate thickness	% wetted surface
Expansion and contraction	Flow configuration
	Tube location
Thermal isolation	Tube bonding to plate
Tube diameter	Turbulent flow

Determining an approach to overcoming the obstacles of collector plate design will largely be a function of material cost and ease of fabrication. In most cases the do-it-yourselfer will have to sacrifice some operating efficiency because of limited access to the appropriate materials. The manufacturer, however, may employ the best possible alternatives which are available. The result is an efficiently designed plate which is often more costly but, in most cases, superior with respect to performance and durability.

Collector Plate Coatings—The plate of a solar collector is black for the obvious reason that black is characteristically absorbent to thermal energy. A black asphalt street or black enamel finish

3: Solar Collection

PHYSICAL PROPERTIES OF SEVERAL METALS

	COEFFICIENT OF THERMAL EXPANSION AT 70°F	SPECIFIC HEAT AT 70°F	DENSITY LBS./FT.^{3}
ALUMINUM	12.90	.224	168.6
COPPER	9.12	.064	558.0
IRON	6.28	.122	492.0
NICKEL	7.20	.111	556.0
TIN	11.90	.054	450.0

FIGURE 20

on an automobile will attest to the high surface temperatures attainable. In a solar collector, however, several new problems arise which will influence the choice of the black absorber plate coating. These factors are listed as follows:

- Shortwave absorptance
- Longwave emittance
- Cost
- Availability
- Life expectancy
- Surface adhesion
- Outgassing

THERMAL ISOLATION OF THE ABSORBER PLATE

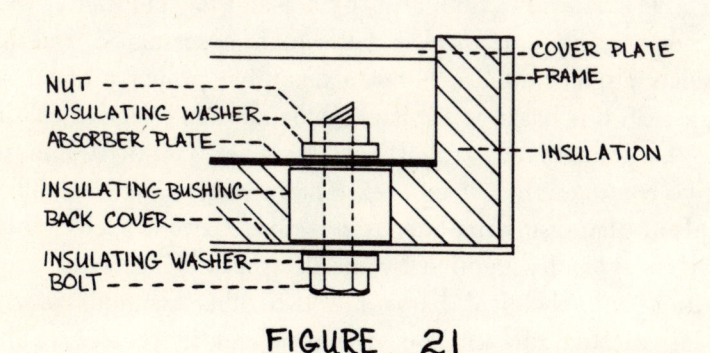

Labels: NUT, INSULATING WASHER, ABSORBER PLATE, INSULATING BUSHING, BACK COVER, INSULATING WASHER, BOLT, COVER PLATE, FRAME, INSULATION

FIGURE 21

53

Solar Energy for the Northeast

THE ELECTROMAGNETIC SPECTRUM

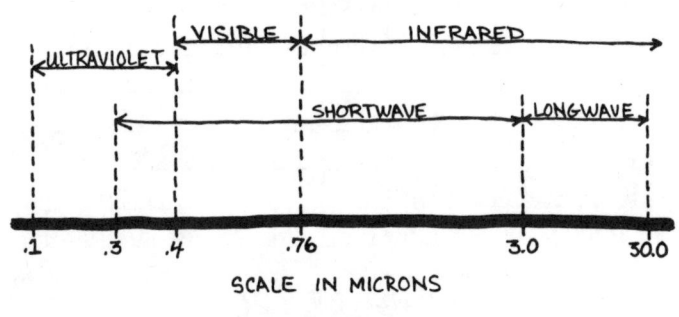

[Diagram showing electromagnetic spectrum with ULTRAVIOLET, VISIBLE, INFRARED regions, and SHORTWAVE, LONGWAVE divisions, scale in microns from .1, .3, .4, .76, 3.0, 30.0]

SCALE IN MICRONS

FIGURE 22

The nature of the radiation incident on the collector surface is illustrated in Figure 22. Incoming radiation is characterized by short wave lengths. When the radiation is absorbed by the surface, it is also transformed into longer wave lengths which are then re-radiated. The purpose of the surface coating is to maximize the shortwave absorptance and minimize longwave emittance. For this reason, Figure 23 is provided to demonstrate the various absorptance and emission coefficients for collector absorber plate coatings.

Surface coatings may be divided into two general categories. A selective surface has high absorptance and low emittance whereas a non-selectve surface has both high absorptance and high emittance. High emittance is not a desirable property of surface coatings. On this basis, it appears as though non-selective coatings should be immediately ruled out. However you will find that selective coatings are not only extremely costly, but also difficult to apply to the absorber plate. A selective surface, such as black chrome, is typically applied by an electroplating process. Vapor deposition and chemical baths are two other methods used in applying selective surfaces. Because of the quality control required by these processes, selective surfaces may cost from 50¢ to $2.50

3: Solar Collection

per square foot of collector. In addition, the home builder will find that the processes are not available commercially unless you wish to plate over 1,000 square feet of surface.

Selective surfaces have also been found to be unnecessary in collectors which utilize two coverplates as a glazing. In this case, the re-emitted energy is trapped in the collector cavity by the insulating barrier of a dual glazing. This is what is commonly known as the greenhouse effect. Although the cover glazing is transparent to the incoming shortwave radiation, it is not transparent to longer wavelengths re-emitted by the plate. Of course, if the plate is cooled sufficiently by the fluid moving through it, then re-emittance of the surface coating will be quite low anyway and the greenhouse effect is negated.

There is also some question as to the durability of selective surfaces. If you consider that you will be plating two dissimilar metals together the question of expansion and contraction arises once again; if the two metals expand and contract at different rates, it is likely that cracking, chipping and peeling may result.

If a dual glazing is used in the collector construction, a non-selective surface may be applied. These surfaces are typically a flat black paint of high carbon content and high temperature

PROPERTIES OF SEVERAL PLATE COATINGS

	ABSORBTIVITY	EMISSIVITY	MAX. AIR TEMP.	DURABILITY
BLACK CHROME	.95	.12	350°F.	15-30 YRS.
BLACK PAINT	.97	.88	350°F.	5-20 YRS.
COPPER OXIDE	.90	.12	200°F.	10-? YRS.
BLACK ALUMINUM	.95	.80	500°F.	50 YRS.

SOURCE: NASA MFS 23518

FIGURE 23

Solar Energy for the Northeast

resistance. Many paints are available for this application. Nextel Black Velvet from 3-M Corporation is in widespread use at this time. It costs around $65.00 per gallon, not including the primer, but its absorptivity and durability are quite high. Other alternatives include wood stove paints, asphalt paints and ordinary flat black paints from the hardware store. Keep in mind that the paint will be exposed to moisture, sunlight (ultraviolet radiation) and high temperatures. The paint must be baked on in the open air to allow gases to escape before the plate is mounted in the collector frame. Failure to do this will result in outgassing and deposition of a thin film on the rear of the glazing assembly. The film will, in turn, decrease the transmissivity of the collector covers. Unless you wish to repaint the plate and clean the glazing periodically, choose the best available alternative. In the event that you do have to re-coat a collector plate, remove the original coat first and polish to a fresh bright surface. The layering of paint on the plate surface will decrease its operating efficiency significantly and you should make every effort to insure a thin uniform coating.

When considering a manufactured collector, do not rule out units with a non-selective surface that have dual glazings. Many manufacturers do not employ a selective surface because its cost

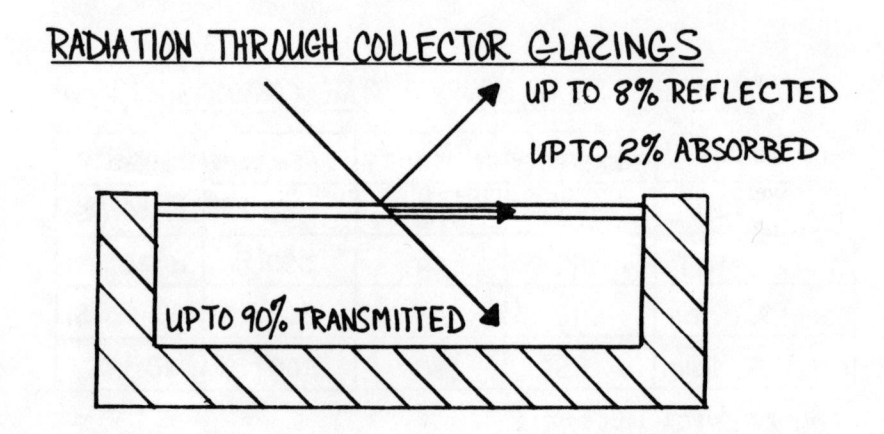

RADIATION THROUGH COLLECTOR GLAZINGS
UP TO 8% REFLECTED
UP TO 2% ABSORBED
UP TO 90% TRANSMITTED

FIGURE 24

PROPERTIES OF SEVERAL GLAZING MATERIALS

	SHORTWAVE TRANSMITTANCE	LONGWAVE TRANSMITTANCE	DURABILITY
WATER WHITE GLASS	91.5	2.0	EXCELLENT
LOW IRON TEMPERED GLASS	87.5	2.0	EXCELLENT
LOW IRON SHEET GLASS	87.5	2.0	EXCELLENT
TEMPERED FLOAT GLASS	84.3	2.0	EXCELLENT
ACRYLIC SHEET	80 TO 89	2.0	GOOD
POLYESTER SHEET	80 TO 85	20.0	FAIR

FIGURE 25

effectiveness and improved performance are questionable.

Glazing Systems—All solar radiation striking the collector plate must first pass through the transparent glazing material or cover plates. The glazing on a solar collector serves the following purposes:

> To transmit radiation to the collector cavity.
>
> To trap radiation within the cavity and minimize heat losses.

As energy strikes the glazing material, it will be affected in three ways. Figure 24 indicates the effect of transmittance, reflectance and absorptance on collector glazing. In addition, there are several other considerations which must be addressed before choosing a suitable cover material. These will include:

> Resistance to abrasion	Ultraviolet degradation
> Resistance to fracture	Longwave transmissivity
> Expansion and	Insulating qualities
> contraction	Cost
> Moisture control

Figure 25 lists the properties of several alternative glazing materials. By and large, glass offers the best results and is the most widely used material. Glass, however, is both expensive and

Solar Energy for the Northeast

subject to breakage. The threat of breakage can be almost entirely eliminated by the use of tempered glass. This is similar to the windshield glass in an automobile with the exception that windshield glass also contains an adhesive film to prevent shattering in the event of an accident. This film is eliminated in tempered solar glass because it would cut down the transmittance through the glass. Very high quality tempered glass has been developed specifically for solar applications. This glass typically has a very low iron content and is termed "water white." By lowering the iron content, the absorptive characteristics of the glass are seriously decreased, leaving a tough, transparent glass with desirable qualities of high transmittance. You can see the iron content of regular glass by observing the green color on the edge.

Solar glass such as "Sunadex" by ASG Industries and "Heliolite" by Combustion Engineering are representative of the highest quality collector glazings available. Unfortunately, it is unlikely that the do-it-yourselfer can acquire this material in small quan-

COMPARISON OF SINGLE AND DUAL GLAZINGS

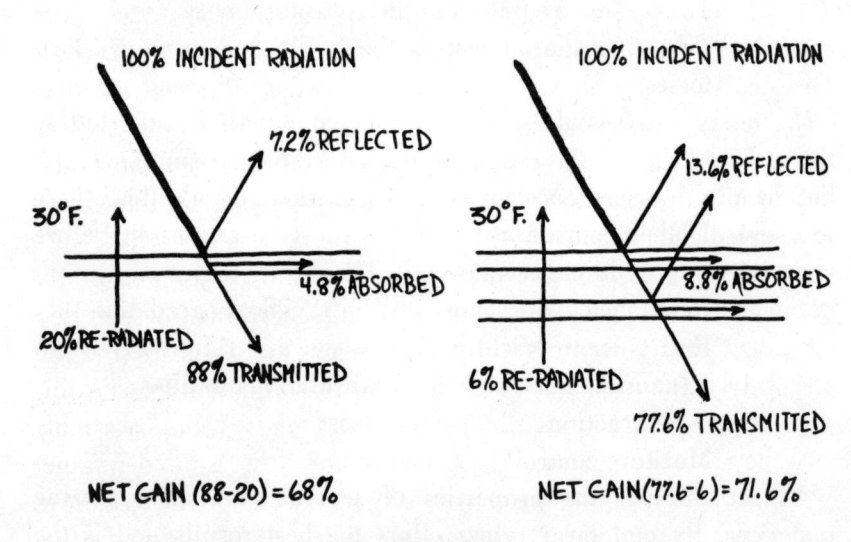

Single glazing: 100% INCIDENT RADIATION, 7.2% REFLECTED, 30°F, 4.8% ABSORBED, 20% RE-RADIATED, 88% TRANSMITTED. NET GAIN (88−20) = 68%

Dual glazing: 100% INCIDENT RADIATION, 13.6% REFLECTED, 30°F, 8.8% ABSORBED, 6% RE-RADIATED, 77.6% TRANSMITTED. NET GAIN (77.6−6) = 71.6%

FIGURE 26

3: Solar Collection

tities. In this event, a suitable plastic or acrylic material must be substituted. Ordinary window glass may be employed as long as thorough consideration has first been given to breakage due to impact or expansion.

Collector glazing plays a major role in controlling heat losses out the front of the collector. In the Northeast, a dual glazing system is required, due to the low ambient air temperature which may be experienced during winter conditions. In southern latitudes, such as Florida or Arizona, the winter temperature seldom drops below freezing and a single glazing is often sufficient.

Figure 26 illustrates the general effect of low ambient temperature on dual and single glazing systems. The air space between a dual glazing acts as an insulating barrier. This air space is most effective when maintained between ½ and 1 inch. The distance of the glazing above the collector plate is not as critical and generally 1 inch will be sufficient.

Care must be taken to insure that moisture does not accumulate and condense between the glazing surfaces. This may be controlled by introducing a desiccant, such as Silica Gel, into the glass spacer. The desiccant will absorb any moisture and prevent water from beading up on the glazing and reducing solar transmittance. Figure 27 gives a typical construction detail to include a desiccant. If the space between the glazings is sealed air tight by silicone or another suitable glass scalant, then only the humidity that was in the air at the time of assembly need be drawn out by the desiccant. No additional moisture can be introduced to a sealed glazing assembly.

Provision must be made for expansion and contraction of the glazing system. The thermal stress on this component of a collector is extreme, as temperatures may reach 400°F on the interior and only 0°F on the exterior. For this reason, a suitable flexible gasket which surrounds the perimeter of the glazing assembly may be included. The gasket is then secured by a fitted retainer which will form a watertight boundary on the face of the collector.

Several anti-reflection coatings are being developed to further increase the transmittance of collector glazings. These coatings

Solar Energy for the Northeast

GLAZING DESSICANT DETAIL

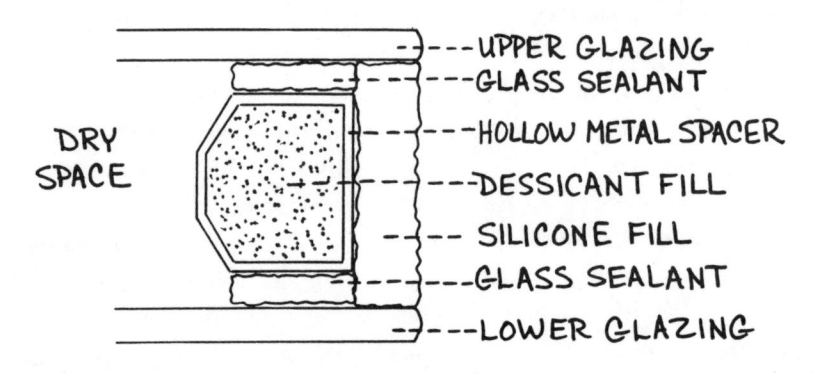

```
                    --- UPPER GLAZING
                    --- GLASS SEALANT
  DRY               --- HOLLOW METAL SPACER
  SPACE             --- DESSICANT FILL
                    --- SILICONE FILL
                    --- GLASS SEALANT
                    --- LOWER GLAZING
```

FIGURE 27

have been noted to achieve up to 95% transmissivity for low iron glass, but they are very expensive and difficult to purchase. There is also some question as to whether they are worth the high cost when performance will only be increased by about 3%.

The most available alternative to glass as a collector glazing is plastic. Many commercial plastics are available to the homeowner at relatively low cost. These include brand names such as: Mylar, Tedlar, Teflon, Lucite, Plexiglass, Lexan, and various polyethylene materials. In addition, a widely used low cost glazing material is Sun-lite, manufactured by the Kalwall Corporation. Sun-lite is a fiberglass reinforced polyester. The plastics are characteristically high transmitting material because the thickness is on the order of a film rather than a plate.

Plastics, however, are also partially transmittant to the longwave radiation being emitted by the collector plate; this means that their effectiveness as an insulating system is lower. In some cases, the increased transmittance may balance out the increased heat loss. In a low temperature pool heating system, this is acceptable, but at higher temperatures, such as those incurred in domestic hot water and space heating systems, the insulating efficiency becomes more critical and glass is recommended.

3: Solar Collection

Plastics are also subject to degradation due to the ultraviolet component of solar radiation. A plastic with a transmittance of 90% may lose up to 10% within one year due to ultraviolet degradation. A yellowing of the material will serve as evidence of this. On some softer plastics, abrasion as a result of tree branches and detritus blowing across the glazing surface can present a problem. A screen can be located over the collector array to prevent damage due to abrasion and impact, although the screen may reduce transmittance by an additional 5%.

In general, plastics are a common alternative as collector glazings. The low cost along with the light weight and serviceability make plastic the most attractive solution in home built systems where the homeowner is prepared to devote some time to periodic maintenance of the glazing system.

High quality manufactured collectors usually employ glass cover plates and thereby eliminate the problems associated with glazing maintenance. The cost of these units also reflects the higher cost of the glass covers.

Insulating Systems—It stands to reason that if you are collecting solar radiation, heat losses in the collector must be kept to a minimum. A poorly insulated collector may lose as much as 40% of its efficiency due to conductive, convective and radiational heat losses. The operating solar collector may be anywhere from 0 to 200°F hotter than the surrounding ambient air temperature. Any cooling that occurs should take place due to the fluid moving heat out of the collector to storage.

In order to insure that any heat lost from the collector is lost to your storage area, a careful examination of the insulating system must be made. We have already examined the collector glazing which acts as the top insulation. The next step is to insure adequate control over the collector sides and bottom. Standard insulating materials will serve this purpose. Two to six inches of fiberglass batt is the most widespread method. Care should be taken that this insulation be at least as thick on the side as on the bottom of the collector. Studies have shown that the average temperature at the edge of the collector plate is often lower than the overall plate due to insufficient side insulation.

Solar Energy for the Northeast

Figure 28 shows a typical collector bottom and side insulation pattern. Note that a reflective foil should be located adjacent to the fiberglass. Above the foil a ½ inch air space allows heat radiating off the plate to be reflected back.

If the collector is to be incorporated into the roof system of the home, any heat losses out the back will enter the living space. Despite this apparent heat gain it makes more sense to insure higher collector efficiency through insulation. If collectors are to be mounted in raised frames on the roof, or ground attached, the frames should be enclosed to minimize heat loss. All piping leading to and from the collector array should also be protected by a standard pipe insulation having an 'R' value of at least 4.0. The 'R' value indicates the material's resistance to heat transfer and is the standard measure used to designate relative thermal qualities of various insulations. Figure 29 gives 'R' values for assorted solar collector insulations.

The type of insulation chosen must be thermally stable at high temperatures (i.e. up to 450°F). Attention should be given to the binder content of fiberglass and its effect on outgassing. Binder is the glue used to hold fiberglass together and fix it to any foil or paper surfaces. At high temperatures this glue can oxidize

COLLECTOR INSULATING SYSTEM

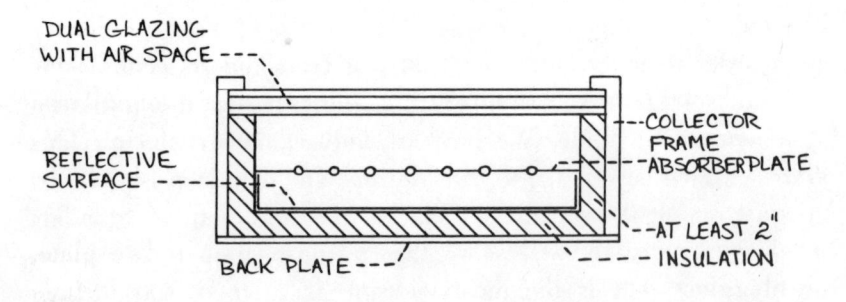

FIGURE 28

and give off a gas which accumulates on the rear of the collector glazing. The residue, in turn, cuts down solar transmittance. Fiberglass having a binder content of less than 1.6 is acceptable.

Polyurethane insulations may deform and give off toxic gases if not chosen properly. Be sure you specify high temperature applications when purchasing urethanes. Polystyrene foams must be limited to low temperature applications where the collector never surpasses 150°F. This practically eliminates polystyrene, as even an inefficient collector may exceed 150°F during stagnation. Stagnation is a period during which no fluid flow is provided

PROPERTIES OF SEVERAL INSULATION MATERIALS

	DENSITY LB./CU.FT.	CONDUCTIVITY (k)	RESISTANCE (R) per inch
GLASS FIBER	4-9	.25	4.0
POLYSTYRENE BEAD	1.0	.28	3.6
POLYURETHANE	2.5	.16	6.2
MINERAL FIBER	15	.29	3.4
WOOD SHAVINGS	8-15	.45	2.2
VERMICULITE	7-8	.47	2.1

FIGURE 29

through the collector. Without a cooling medium the collector will slowly build up temperature and press all of its components to their respective high temperature limits. Consideration must also be given to the flammability of an insulating material. This data is generally provided by the manufacturer and should be available on request.

Collector insulation is not expensive when compared to plate, glazing and frame costs. It makes good sense then to include adequate insulation when constructing the unit; its ease of handling and availability make it one area of collector design where the best possible alternative is available to the homebuilder.

Solar Energy for the Northeast

Collector Frame—The collector frame may consist of aluminum, galvanized iron, fiberglass, plastic or wood. The differences between these materials are readily apparent. A wood frame, unless properly treated and maintained, will eventually rot, or dry out and split. This will require replacement of the entire collector as its usefulness will be marginal with a gaping hole in its side. Plastic and fiberglass frames must be able to withstand the constant heating and cooling cycle without distressing and cracking. Few plastic frames will endure for twenty years without serious deformation. Galvanized iron frames are durable but heavy.

Aluminum represents the best alternative in frame design. It is light, durable and readily available in sheet, plate and simple extruded forms. Manufactured collectors generally use an extruded aluminum frame although in many cases special dies are required to form more versatile and complex extrusions. The aluminum frame must include adequate provisions for mounting the glazing system, the collector plate and the insulation. The frame should be helio-arc welded to insure a tight continuous enclosure. It must also be watertight to prevent moisture from entering the collector and corroding the collector plate or insulation. Mill finish aluminum will wear nicely when exposed to the elements and deterioration will be practically non-existent, though some pitting may occur in shoreline areas where salt spray is carried in the air. If necessary, the aluminum can be primed with zinc chromate and painted with a high quality enamel to provide a more attractive finish.

In new construction, the roof rafters can act as the collector frame. They will not be exposed to the elements, so deterioration is no longer a problem. The collector plate must then be attached between the roof rafters and a suitable glazing system incorporated in the top. Fiberglass insulation may be added to the bottom as in a conventional roof. The roof-integrated solar collector requires careful planning and installation to prevent heat losses and moisture buildup. In many cases, however, this represents the most cost-effective alternative. The expense of additional framing for individual collectors is eliminated. The roof integrated frame

3: Solar Collection

ROOF INTEGRATED AIR COLLECTOR

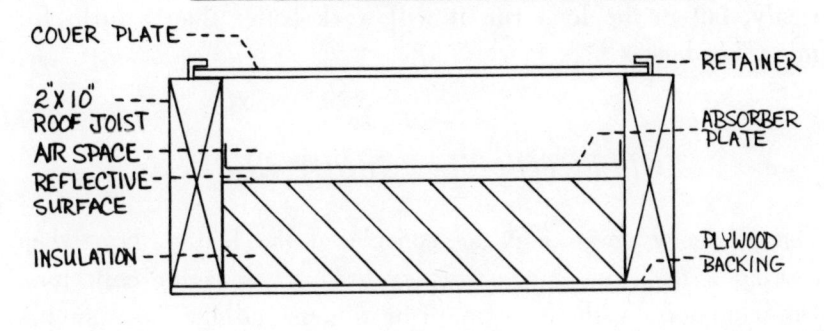

Labels: COVER PLATE, RETAINER, 2"X10" ROOF JOIST, ABSORBER PLATE, AIR SPACE, REFLECTIVE SURFACE, INSULATION, PLYWOOD BACKING

FIGURE 30

is ideal for air collection systems where no piping is required and collector plates are simple by design.

Figure 30 shows a possible design for solar collectors which are located between 2" x 10" rafters spaced 24" on center.

Gaskets and Seals—Several properties must be taken into consideration when choosing gasket and sealing materials. These include:

    Ultraviolet resistance
    Resistance to -20 to $450°F$ temperature cycling
    Surface adhesion
    Sufficient elasticity and compressivity to withstand expansion and contraction
    Weatherability

The materials appropriate for use in solar collectors are from a general category called elastomers. Currently many gaskets are made of EPDM (ethylene propylene diene monomer) rubber; this will withstand temperatures up to $350°F$ without degradation. Seals may also be EPDM or a silicone type rubber. The silicone sealants are expensive, but superior in both weathering and high temperature properties. The upper limit on this material is $450°F$. In all cases, apply a degreasing solvent to the surface prior to sealing with silicone.

Solar Energy for the Northeast

The proper choice in gaskets and seals will substantially reduce maintenance requirements. At the onset, this will appear more costly, but in the long run it will work better than aspirin for minor headaches.

COLLECTOR EFFICIENCY CURVES

The efficiency curve may be considered the bottom line when making a decision to construct or to purchase solar collectors. Unfortunately, without some prior understanding, these simple curves can be both evasive and deceptive. The conditions under which the curve is derived are sometimes either unclear or unknown to the salesman who is ready to bet his life on their accuracy. A common mistake is to evaluate the curve only from the vertical axis. It is this axis which indicates the percent efficiency of the tested unit. The curve will intercept the vertical axis at a point which is known as the peak efficiency. The peak efficiency is not the efficiency at which the collector will operate once it is up on your roof. Remember this and you will most likely be ahead of the salesman right from the start. In order to stay one step ahead, you should also be familiar with some basic thermodynamics of solar collectors.

The useful heat output of a collector is equal to the amount of energy falling upon the collector absorber plate minus the total heat losses of the unit. It can be extremely difficult to summarize all the rates of heat loss for varying conditions. For this reason, it becomes more practical to measure collector output by measuring its effect on the fluid which is passing through it. This can be easily accomplished by locating thermometers on the inlet and outlet connections. The temperature leaving the collector minus the temperature entering is known as the delta T ($\triangle T$) or collector differential. The higher the differential, the more efficient the unit is at capturing radiation. Most manufactured units have a differential of between 10 and 18°F under peak operating conditions. Of course, the performance of the collector will vary

3: Solar Collection

depending on whether it is in San Antonio, Texas or Bangor, Maine. For this reason, collector efficiency curves become somewhat less useful when trying to determine the output at your site. However, the curves are quite capable of demonstrating the relative performance of two different manufactured collectors. This situation is analogous to the E.P.A. mileage estimates which are now required of all automobile manufacturers. Although the E.P.A. estimate states that your brand new car will get 26 miles to the gallon, you know that it actually has never topped 20 miles to the gallon. The E.P.A. goal is not as much the accurate prediction of your mileage as it is a useful comparison of how your vehicle stacks up against others on the market under identical conditions.

The E.P.A. does not test solar collectors, but several other reputable organizations do and standardized test results should be attainable from any manufacturer. The most advanced testing program to date has been implemented by ASHRAE. These are the same people that provide similar data on all conventional heating and cooling equipment. The S.E.I.A. (Solar Energy Industries Association) is presently effecting a similar program, but the program is far from complete and currently many manu-

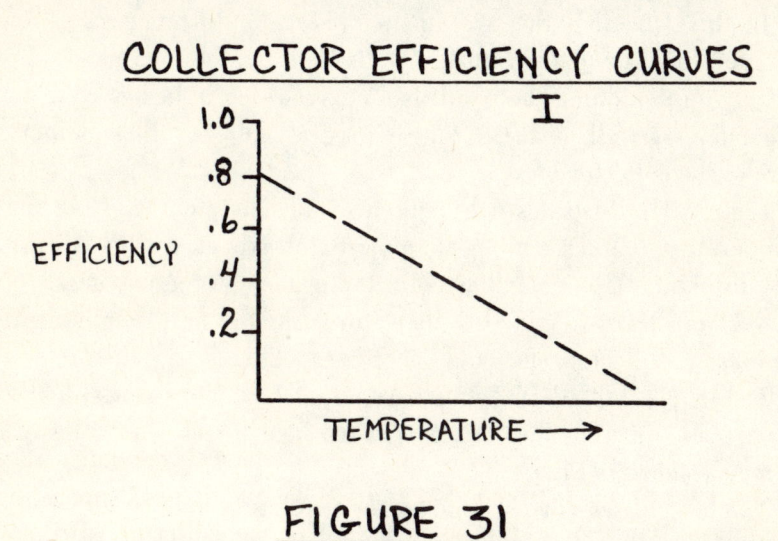

COLLECTOR EFFICIENCY CURVES

FIGURE 31

Solar Energy for the Northeast

facturers are not included in the S.E.I.A. listings. The D.O.E. (Federal Department of Energy) has also tested more than 300 collectors currently available but the data has yet to be released to the public.

By far, the ASHRAE data is the most reliable and available. ASHRAE developed the test procedure and many laboratories throughout the Northeast are capable of applying it. This testing is required of all manufacturers who are involved in the H.U.D. grant program as well as similar Federal and state funded projects. In short, any reputable manufacturer should have the data on hand. Curves which are provided as a result of the manufacturer's own test program may be questionable.

In ASHRAE curves, collector efficiency is plotted against temperature differential. The differential function is more precise than the simple T outlet minus T inlet discussed earlier. It is, in fact, the specific temperature difference or:

$$\frac{\frac{(T\ inlet\ +\ T\ outlet)}{2} - T\ ambient}{Total\ Insolation}$$

In this function, the average temperature difference, less the surrounding air temperature, is divided by the total radiation falling on the plate. The result is a curve which indicates collector efficiency, as well as relative heat loss, as the operating temperature of the collector increases.

Figure 31 illustrates a hypothetical efficiency curve. Note that temperature increases along the horizontal axis as you move from left to right. Also note that collector efficiency decreases as you move from left to right. In other words, as the fluid temperature increases, collector efficiency decreases. This is because as the collector gets hotter, its tendency to lose heat increases. Ideally, a perfectly insulated collector without heat losses would have a curve as shown in Figure 32.

The heat losses (or conversely, the insulating quality of a solar collector) may be interpreted as the slope of the efficiency curve.

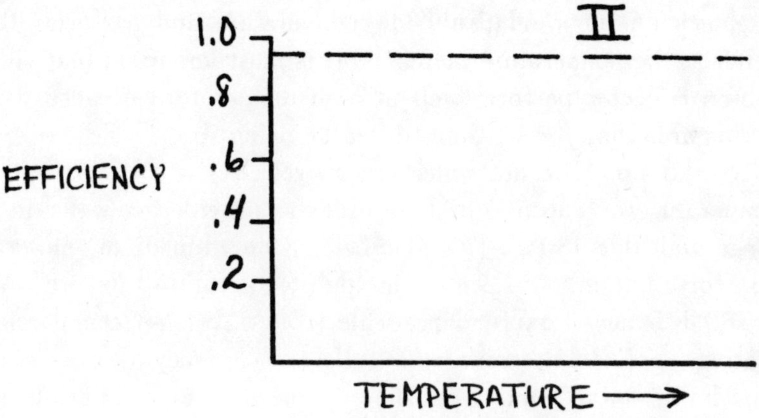

FIGURE 32

This is quite likely the most valuable tool in evaluating performance data. It allows you to distinguish between a collector with high baseline efficiency and high heat losses as opposed to a unit of slightly lower initial efficiency but sustained performance at high temperatures. Figure 33 is representative of this comparison.

Notice how collector 'A' starts out at 80% efficiency and presumably "out-performs" collector 'B'. Collector 'B', however,

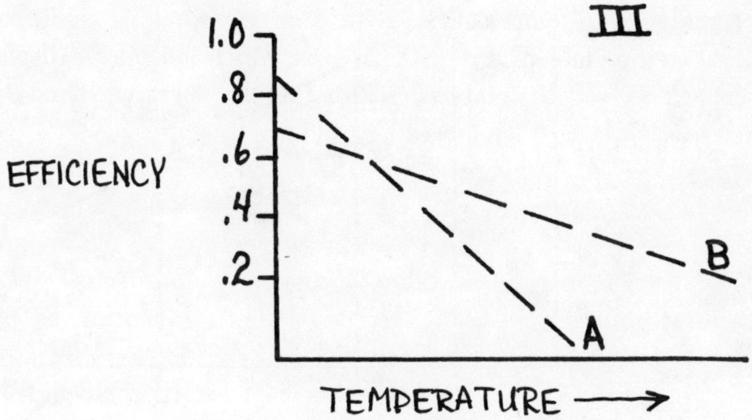

FIGURE 33

Solar Energy for the Northeast

intersects 'A' at a relatively low temperature and performs far better as the temperature increases. It is most important that your chosen collector perform well at high temperature because it is in this area that you will most often be using it.

A solar domestic hot water system requires a collector outlet temperature of 130 to 140°F in order to provide hot water in a range of 120 to 130°F. This may be considered medium temperature for a flat-plate collector. The high temperature range of 160 to 180°F is necessary for large-scale (over 125 tons) commercial cooling applications. Since the collector efficiency decreases at this level, solar cooling becomes somewhat more difficult to accomplish. Home air conditioning systems require from 190 to 210°F water to drive absorption chiller units. At this temperature the collector's efficiency is seriously decreased and the feasibility of solar home cooling is practically ruled out. Figure 34 summarizes solar applications as a function of collector efficiency.

EFFICIENCIES OF SOLAR APPLICATIONS

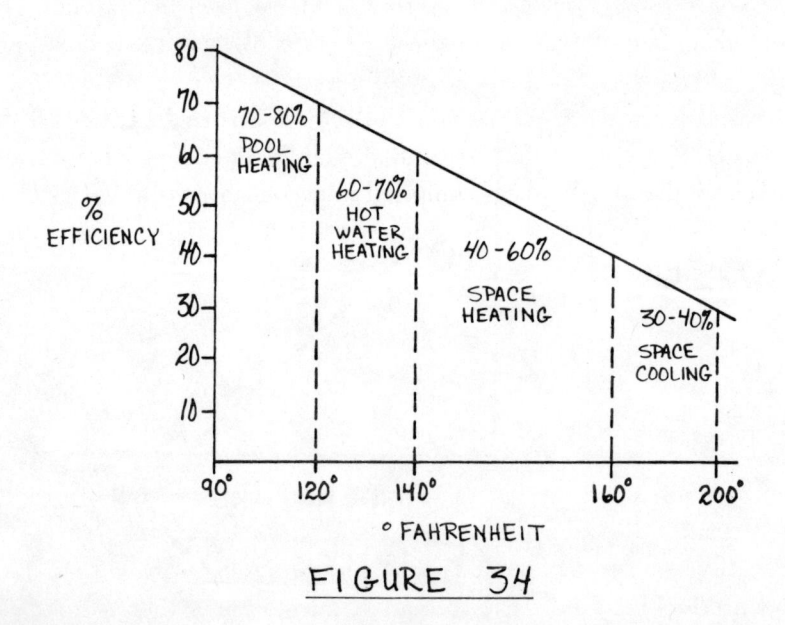

FIGURE 34

3: Solar Collection

Pool heating is all the way down at the opposite end of the scale. At temperatures of 80 to 90°F, the collector efficiency is very close to the maximum attainable. As a result a low cost or even a home built system will often perform quite adequately.

One other important aspect in collector performance is the temperature of the environment it is operating in. This ambient air temperature may vary from around 0 to 100°F in New England. At 0°F the collector will have a greater tendency to lose heat than at 90°F, due to the larger temperature differential. Because of this, the solar collector often performs somewhat less efficiently during the winter months. Again the insulating quality, or the slope value from the performance curve, is important to minimize this effect.

As you can see, the efficiency of a flat-plate solar collector is a transient function, subject to several variables. In no case may a firm % efficiency figure be quoted to you on collector performance. With this knowledge, you can question a manufacturer and determine how well he understands his own business. If you can stay ahead of him all the way through your discussion, then it is probably advisable to consider another manufacturer. There is, of course, the possibility that in a new and rapidly expanding field, such as solar energy, the sales personnel may not be trained to handle detailed queries on collector performance. In this case, obtain a copy of the test results by an independent laboratory and check that the data was compiled in accordance with ASHRAE procedures. Perhaps, at this point in solar applications, the rule is trust yourself, not the salesman.

COLLECTOR PERFORMANCE

The question repeatedly comes up as to what exactly will the solar collector put out in terms of useful energy. Despite the volumes of literature and pages of manufacturers' specifications, you will probably not be able to find out this one simple fact. The data you are looking for is Btu output. It may be on either a square foot or gross collector basis and it will probably be per hour or day.

To determine the Btu output of a flat-plate solar collector, you will need to be familiar with three variables. These variables change from manufacturer to manufacturer and are generally available on request if not already in the sales literature. They are:

Fluid mass flow rate: The fluid moving through a collector will flow at a constant rate which is given in gallons per minute (gpm). All collectors have a recommended flow rate in order to obtain optimum conditions for heat transfer. Often this flow rate will vary between .25 and .50 gpm. In order to determine the mass flow rate, you will need to know the weight in pounds of a gallon of the recommended fluid. This will then yield pounds per minute per collector. From this point you can easily convert to pounds per hour per square foot of collector.

For example, if a manufacturer uses water as a heat transfer fluid and the recommended flow rate is .50 gpm, then you can find that 1 gallon of water weighs 8.3 pounds and that .5 gallons weighs 4.15 pounds. You now have a mass flow rate of 4.15 pounds per minute or 249 pounds per hour. This is the flow rate for the entire collector. To determine flow rate per square foot, simply divide the mass flow rate (249 lbs/hr) by the area of the absorber plate. For this example, assume the collector absorber plate is 18 square feet, divide this into 249 lbs/hr to yield 13.8 lbs/hr/square foot of collector.

Once you have found the mass flow rate in lbs/hr/sq. ft., you know approximately how much fluid is moving through the collector and now must find the ability of that fluid to absorb heat.

Specific Heat of the Fluid: The specific heat of a fluid indicates that fluid's ability to absorb energy. For water, the specific heat is 1.0, and for air it is .24. Many fluids are currently being employed which have specific heats varying between .30 and .70. A detailed list of these is presented in the section on heat transfer fluids. It is reasonable to assume that you would wish to choose a fluid with as great an ability to absorb heat as possible. For this

3: Solar Collection

reason water seems the appropriate alternative. However, due to problems with temperature range and corrosion, water may not be the best alternative. The specific heat is given in Btu/lb/°F and is, again, available from the manufacturer.

Knowing the specific heat allows you to find the amount of energy capable of being absorbed by the fluid regardless of how fast the fluid moves through the collector. Coupling this data with the flow rate you will now require only one more step to determine Btu output.

Collector Temperature Differential: This factor is undoubtedly the most difficult to pin down. It will vary from location to location as solar availability varies. It will also change as the ambient air temperature changes and, most important of all, it will be a function of the temperature range in which the collector is operating.

The temperature differential is obtained by subtracting the collector inlet from the collector outlet temperature in °F. Because the collector will perform differently under various environmental and operating conditions, the most you can hope for is an average differential. Most manufactured collectors have a differential in the area of 12 to 22°F. Obviously the higher the difference the more heat the collector is capturing and the more efficient the unit. At low inlet temperatures (i.e. 80°F), the differential will be highest if you recall that efficiency is greatest at low temperature. Differences as high as 30°F may be achieved in well-built units. For medium temperature (i.e. 130°F) applications in the New England area (such as domestic water and space heating) the differential is more likely to be around 10 to 15°F. The manufacturer should be able to tell what temperature differential to expect in your area for your application.

With an understanding of mass flow rate, specific heat and temperature differential, you are now prepared to carry out a simple calculation which will determine the useful energy collected. Recall that this is the energy contained in the fluid and does not take into consideration heat losses that occur in transit or in storage.

Solar Energy for the Northeast

The equation is as follows: Mass flow rate × specific heat × temperature differential = total collected energy. As long as you can keep the units straight, the result will indicate the Btu output of a collector. To review the units the following equation is given:

$$\frac{\text{LBS.}}{\text{SQ.FT.-HR}} \times \frac{\text{BTU.}}{\text{LB.-°F}} \times °F = \frac{\text{BTU}}{\text{SQ.FT.-HR.}}$$

By example, we will refer back to a simple system utilizing water as a heat transfer fluid with a flow rate of .5 gpm in an 18 sq. ft. collector. The manufacturer states that a temperature differential of 15°F may be expected under peak conditions. With this data the equation becomes:

$$\frac{13.8 \text{ LBS.}}{\text{SQ.FT.-HR}} \times \frac{1.0 \text{ BTU.}}{\text{LB.-°F}} \times 15°F = \frac{207 \text{ BTU.}}{\text{SQ.FT.-HR.}}$$

The above-described collector will yield 207 Btus per square foot per hour. If the collector was 18 square feet, then this becomes 3,726 Btu per collector per hour. If you receive 6 hours of sunshine per day then the daily output of the theoretical solar collector is 22,356 Btus!

Unfortunately, most collectors do not use water as the fluid media due to freezing, boiling and corrosion problems. In this case, the specific heat factor of the equation may be reduced by as much as one-half with the resultant effect of the collector yielding around 11,000 Btus per day. Even at 11,000 Btus per day, a 3 collector system will easily deliver the 30,000 Btus per day required to provide domestic hot water for a family of four.

This equation is useful in that it brings home the point that flat-plate solar collectors will yield substantial amounts of energy for domestic uses. In addition, it is useful if you want to check on the performance of a system already installed or if you are judging the reliability of a manufacturer with whom you may do business. In either case, this steady state equation of the thermal performance of solar collectors is basic to an understanding of the operating parameters involved in a flat-plate collector system.

HEAT TRANSFER FLUIDS

Fluids used in solar collection systems are quickly becoming prime considerations in the design phase of solar equipment. Initially, either water or water/glycol solutions were considered until the effects of these media was fully determined. Currently glycols, aromatic oils, parrafinic oils, silicone oils and synthetic hydrocarbons are available to meet fluid requirements.

The choice of a heat transfer fluid will be based primarily on the following points:

Toxicity: If a fluid is to be used in a system that will provide domestic hot water, then it is likely that a heat exchanger will be involved to transfer heat from the fluid to the hot water supply. This hot water supply is termed potable; in other words, because you use it to shower and to wash dishes, it must not be harmful if ingested. Admittedly, most of us do not drink our shower or dishwater, but nonetheless it must be pure and uncontaminated as it comes in contact with us. If a leak develops in a heat exchanger, then the opportunity arises for the heat transfer fluid to move into the water supply. If this fluid is toxic, then the water supply becomes contaminated and a hazard may exist. Double wall heat exchangers may be used to prevent fluid leakage into the water supply but these will also cut down the heat transfer capabilities. Many states now specify that these exchangers must be used if a toxic fluid is in the system. As a warning device, most toxic fluids have a colored dye included which will show up at the faucet if a leak occurs. In this case, it is important that a bypass loop has been installed on the solar system in order to isolate it from the remaining water supply.

Toxicity, in general, is not a major problem as most fluids are either non-toxic or only mildly toxic. In many cases, the most severe effect of ingestion may be a brief period of diarrhea. However, it is worth the time to question a manufacturer on this point.

Flammability: The flammability of a fluid can be most clearly determined by its flash point. This is the point at which the fluid vapors will burn if an ignition source is present. The flash point should always exceed the system stagnation temperature; for most systems 400 to 450°F is considered safe.

Thermal Stability: The fluid must be able to withstand high temperatures without decomposing into potentially corrosive acids. For some fluids, the presence of metals such as copper will aid in this decomposition. As a fluid decomposes, its viscosity may also increase and cause pumping problems due to greater frictional losses. In addition, scale may be deposited on the interior surfaces of the absorber plate and decrease the rate of heat transfer.

Freezing Point: A system operating in New England must be capable of withstanding temperatures well below 0°F; this is the primary reason that water has been eliminated as a choice. Although there are some who would argue for automatic draindown systems, the potential for trouble is there. Consider what happens if a solenoid valve sticks and fails to drain the collectors or a power failure causes the valve not to be actuated. There is no easy way to replace an entire collector array which has frozen and ruptured. A freeze inhibitor must be added to liquid systems or the fluid itself must remain liquid down to −20°F. Many fluids are available which meet these requirements adequately.

Specific Heat and Thermal Conductivity: The fluid is the major component in the thermal circuit which transfers energy from collector to storage; its ability to absorb and conduct energy are prerequisites to efficient performance. A high specific heat indicates a greater capacity to absorb energy.

Life Cycle and Maintenance: Depending on the oxidation rate, the fluid may have to be replaced periodically. Just as you replace

3: Solar Collection

antifreeze in your automobile, similar glycol solutions need servicing in a solar system. In most closed loop systems the air is entirely eliminated and therefore oxidation is inhibited for oil-type fluids.

Costs: Fluids may range from $1.25/gallon for weak glycol solutions to more than $25.00/gallon for sophisticated silicone oil transfer systems. The cost of the fluid is closely related to its maintenance and replacement schedule. The silicone fluids will not degrade when installed properly and therefore do not require replacement for the life of the system. This can offset high initial costs as well as decrease overall maintenance requirements.

PROPERTIES OF SEVERAL HEAT TRANSFER FLUIDS

	COST $/GAL.	SPECIFIC HEAT	FREEZING POINT °F	TOXICITY
WATER		1.00	32	
ETHYLENE GLYCOL	2 - 3.	.70	-50	HIGH
PROPYLENE GLYCOL	2 - 3.	.70	-50	LOW
SILICONE OIL	20. - 30.	.37	-80	LOW
PARRAFINIC OIL	3. - 4.	.45 - .55	-20	LOW
AROMATIC OIL	3. - 4.	.45 - .55	-20	MEDIUM

FIGURE 35

Figure 35 indicates the properties of several widely used heat transfer fluids. Each fluid has advantages and disadvantages which must be thoroughly considered. Often, disadvantages can be overcome by adjustments in system design, although these changes generally increase the total cost of the system.

There are four principal heat transfer fluids in use today. No particular fluid may be recommended as this will depend to some extent on the type of system being considered.

Solar Energy for the Northeast

Water: The use of water in solar systems immediately implies that a draindown configuration will be employed. In some cases, distilled water with freeze and corrosion inhibitors will be specified. Once these inhibitors are added, the fluid may no longer be non-toxic and the appropriate steps must be taken to prevent contamination of potable water supplies. Water has the advantage of a high specific heat and low cost. At the same time, its disadvantages include relatively high freezing and low boiling points, as well as problems associated with galvanic corrosion and scale formation.

Propylene Glycol/Water: Glycol solutions typically can withstand freezing temperatures, but still have a problem at the upper end of the temperature scale with boiling. During stagnation a boiling glycol solution will decompose to form organic acids and sludge which will, in turn, reduce collector efficiency. Sludge also creates pumping problems if the fluid is not replaced periodically. A properly operating glycol solar system will require that the homeowner accept responsibility in maintaining the fluid at the manufacturer's recommended pH. This will control the acid/alkali balance and insure that undue deterioration in the system components does not occur. The use of propylene in place of ethylene glycol is recommended to minimize toxicity. The glycols are relatively low priced and easily accessible to the homeowner. Their accessibility, however, does not reflect on their disposability. Since their maintenance requires periodic changing of the fluid the problem arises as to how to get rid of the waste.

Silicone Oils: The silicones represent a truly valuable accomplishment from the chemist's perspective. These oils are essentially non-toxic and resistant to chemical degradation. No maintenance will be required in a silicone transfer system. In addition, these oils will not freeze or boil or support corrosion in collector absorber plates. There are, of course, several drawbacks to silicone which include their high cost and poor transfer characteristics. The problem with heat transfer capability arises out of the nature of the fluid and its flow characteristics. As discussed earlier, a fluid which flows turbulently will mix and heat evenly thereby

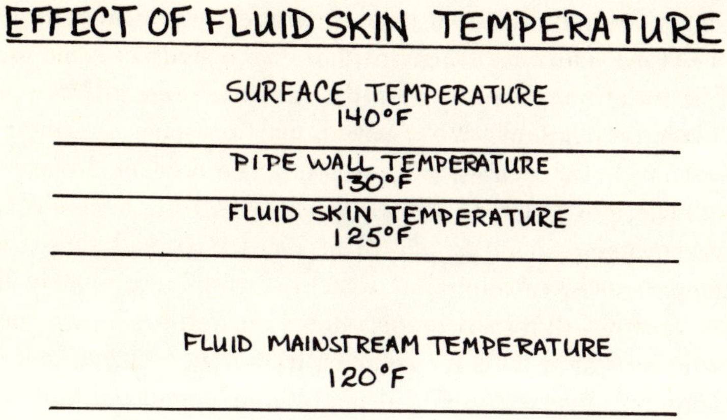

3: *Solar Collection*

EFFECT OF FLUID SKIN TEMPERATURE

SURFACE TEMPERATURE
140°F

PIPE WALL TEMPERATURE
130°F

FLUID SKIN TEMPERATURE
125°F

FLUID MAINSTREAM TEMPERATURE
120°F

FIGURE 36

increasing its ability to absorb energy. A fluid, such as silicone, which flows in a laminar fashion will develop a high temperature at the pipe/fluid interface but a relatively lower temperature in the mainstream. The mainstream flows somewhat faster as it is not in contact with the piping wall. Figure 36 illustrates this effect. In addition, some question remains as to seepage in soldered plumbing joints. Although the manufacturer usually cites poor workmanship as the primary cause of leaks, there have been cases where experienced plumbing fitters have been unable to contain the fluid.

Hydrocarbon Oils: These oils are similar in nature to hydraulic fluids and have been used extensively as heat transfer fluids in the power generating industry. Generally, the oil is used to cool large scale transformers and prevent the problems which could result from overheating. This is similar to the oil's function in a solar collector. Commonly used oils in solar systems are a paraffinic base with the unstable components separated out. In this form, hydrocarbon oils are relatively inexpensive and non-toxic. The oil must be resistant to thermal cracking at high temperatures. Thermal cracking is the breakdown of a fluid into thick tar substances or coke. The coke in turn may deposit on the

Solar Energy for the Northeast

absorber plate and create an insulating effect which impairs heat transfer. Care must be taken in choosing a hydrocarbon oil to insure that these undesirable qualities are not present. Many oils are currently available with exceptional resistance to thermal degradation. In most cases, these oils must be used in closed loop systems where exposure to air is prevented. Attention should also be given to pump sizing as the oil becomes thick and sluggish at low temperatures, causing slow system start-up. Fortunately, the solar collectors will minimize this effect in a well-designed system.

When choosing a fluid for solar collection systems, the options are many and diverse. A careful analysis of the above fluids will serve as a starting ground from which to make a decision. Cost, maintenance, stability and thermal performance will be the expected tradeoffs.

4: SOLAR STORAGE

One of the questions most frequently asked of solar engineers and architects is, "What do I do when the sun isn't shining?" It is true that heating requirements begin to increase as soon as the sun goes down, and this fact in itself has led many people to realize the tremendous amount of energy which the sun provides during its daily traverse from east to west. Were the sun available on a 24 hour basis, a well designed passive structure might easily provide all its own heating requirements. As it stands, however, there must be a means of saving that daytime energy for later use.

The sunshine which is available during the day exceeds the house's daytime heating requirement; any passive solar design will demonstrate this effect quite well. Conventional backup systems are usually not required until later in the evening and early morning hours. Since more energy than you can possibly use is available during the daytime, the remainder may be diverted to storage where it can be regulated and distributed as the need arises.

The means of energy storage raises a topic which is currently in the forefront of solar design discussions. The reason is that solar energy can be both awkward and difficult to contain. The volumes required are typically quite large and therefore prone to problems associated with cost, space and performance. Unlike a gallon of oil which contains 138,000 Btus of energy and may be stored almost indefinitely, a similar amount of solar Btus may require 1,000 gallons of water stored at 70°F with a constant tendency to cool and lose energy. There is a simple reason for this dramatic difference: where a gallon of oil contains potential energy, a tank of solar hot water contains kinetic energy. Kinetic energy, often termed "energy in motion," is released energy. In the solar case, it is energy released from the fusion of hydrogen atoms 93 million miles away. This energy, once released, can be contained again about as easily as energy released from burning a gallon of oil. The most you can do is use the heat to warm a media such as water or rock and then keep that media from losing its heat by careful insulation. Although a difficult task, solar heat storage can be accomplished effectively with available materials. Many strategies have been employed with varying degrees of success. Each of these strategies may be considered to fall into one of the following categories:

Latent Heat Storage: utilizing phase changing eutectic salts

Sensible Heat Storage: utilizing water or crushed rock

Latent Heat Storage through the use of eutectic salts is an emerging science which has yet to prove itself. Its potential is great because the associated costs are low as well as the volume of material required. Theoretically, the operation of a eutectic salt storage system is quite simple. The most commonly used material is known as Glaubers salt or sodium sulfate decahydrate. This salt melts at 90°F and as it liquifies stores 105 Btu per pound. Later, when the melted salt re-solidifies, it gives off the stored energy for use in space heating. As you can see from performing a quick calculation, the amount of salt required to store 20,000 Btus of energy is only 190 pounds. The space required for this storage system is equally small when compared to water or rock.

Several systems utilizing phase changing storage are now on the commercial market. By and large these systems are unproven and somewhat risky. Among the drawbacks:

4: Solar Storage

1. The salt must be able to withstand many phase changing cycles without losing its latent heat storage potential. This has been a point of focus for years of research. Various inhibitors have been added which prevent the salt from separating out into its various constituents.
2. The phase change must occur within a relatively small range close to its actual melting temperature.
3. The surface area of the containers must be maximized to insure a complete and uniform change of state throughout the entire media.
4. It must be non-toxic, non-corrosive and non-combustible.
5. It must be cost-effective with low maintenance requirements.

Most manufacturers will insist that these drawbacks have been overcome successfully. Unfortunately, this author has not had the opportunity to extensively test these units for their reliability and is therefore unable to make a recommendation at this time. The phase changing salt storage systems do bear further investigation, however, as their apparent capabilities are significantly outstanding.

Sensible Heat Storage through the use of water or rock is by far the most widely used means of solar storage. Utilized in thousands of residential and industrial applications throughout the Northeast and the U.S., these two media have been extensively tested in working systems.

The capacity of a sensible heat system is a function of the specific heat of the storage medium. As discussed earlier, the specific heat is the number of Btu's required to raise one pound of the material by 1°F. As the material cools it gives off the stored heat for useful purposes.

A common question is whether to use rock or water and how much to use. Water has a specific heat of 1.00 Btu/lb/°F whereas stone—may be as low as 0.21 Btu/lb/°F. In this case it is obvious that you can use a smaller mass of water in place of stone. The stone will require up to five times the mass. On the other hand, stone is more easily contained than water. It can also be used more effectively in air type collectors without attention to heat exchangers, piping and fluid leakage. In effect, a rock storage system is simple, easily constructed by the homeowner, and relatively little maintenance is required. In addition, the rocks will never wear out or leak.

From the design perspective however, a water storage system is more versatile and takes up less space. It is also the most effective storage media for solar hot water systems. It doesn't make sense to heat a bed of rocks when your end goal is hot water.

Regardless of your choice of heat storage media, you can count on one fact: the volume will be large. This poses a serious design question to the architect or builder. For existing homes, the addition of large storage areas is virtually impossible unless you are prepared to flood the basement or dump several truckloads of gravel through the cellar window.

To get an idea of the volume required, consider 50 to 300 cubic feet to store 500 to 2,000 gallons of water, or 150 to 800 cubic feet to store 10 to 50 tons of stone. It is obvious that these areas should be incorporated into the structure during the design phase if they are to result in an unobtrusive and harmonious building facade.

Figure 37 demonstrates several alternative locations for the thermal storage mass. In all cases, it is to the designer's advantage to increase the overall mass of the structure. Buildings incorporating high mass will change temperature more slowly. A wood frame ranch home may vary between 40 and 70°F during the course of a day whereas a primarily stone or concrete structure with large thick interior walls may vary between only 50 to 60°F. The goal is for the building to remain resistant to major outside temperature fluctuations. This has been most effectively accomplished through locating the building substantially below ground level or into a hillside; the surrounding earth acts as a buffer to extremes in outside air temperature. In most cases, however, the homeowner will prefer to locate the home in a conventional man-

ALTERNATIVE STORAGE LOCATIONS

EXTERIOR WALL

BASEMENT CRAWL SPACE

INTERIOR WALL

ATTIC SPACE

ABOVE GROUND

BELOW GROUND

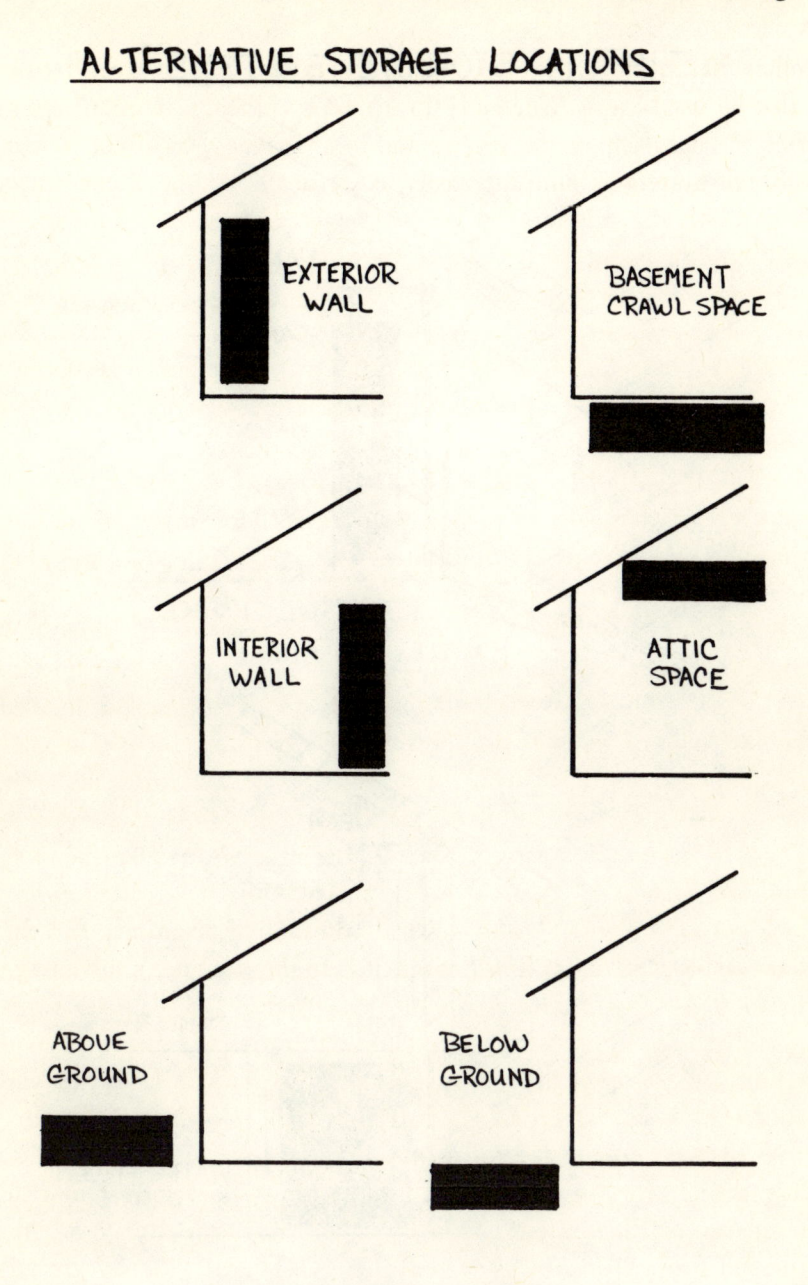

FIGURE 37

Solar Energy for the Northeast

SIMPLE STORAGE SYSTEMS

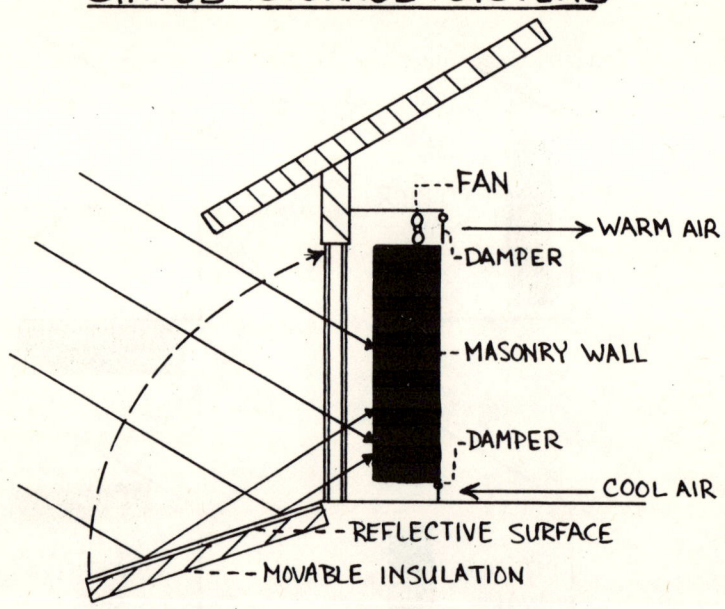

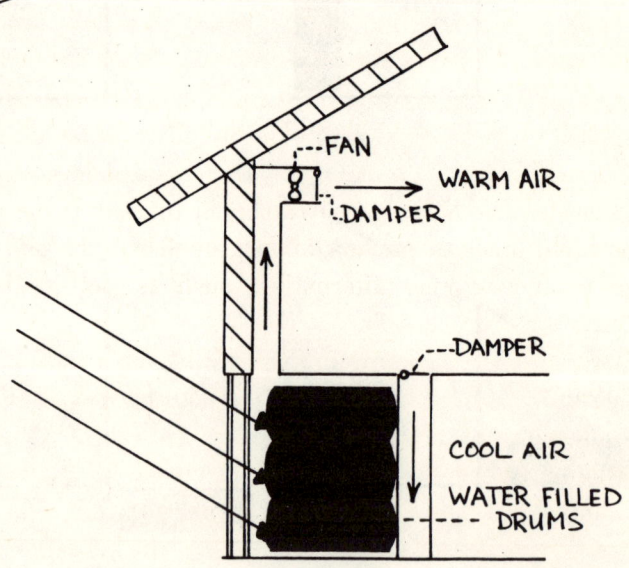

FAN
WARM AIR
DAMPER
MASONRY WALL
DAMPER
COOL AIR
REFLECTIVE SURFACE
MOVABLE INSULATION

FAN
WARM AIR
DAMPER
DAMPER
COOL AIR
WATER FILLED DRUMS

FIGURE 38

86

4: Solar Storage

ner. The storage concepts illustrated offer the possibilities that are available.

When locating storage mass on the building site, several additional factors should be taken into consideration. To operate a system efficiently, heat losses occurring in transit from collectors to storage should be minimized. In other words, locate the storage area near the collectors. If possible, also consider placing the storage mass where it will be exposed to passive solar collection as well as active.

If you choose to place the storage area inside the structure, the basement is the most obvious alternative. Consider the additional loading on foundations and footings which will result. A rock bin weighing 50 tons or a water tank weighing 10 tons is considerably more than the normal load anticipated. In some areas of coastal New England, the load bearing capacity of the soil must also be tested or you may end up with an underground home despite your aesthetic preferences.

When choosing outside storage areas, remember that any heat losses which occur will be true losses. Unlike interior storage areas, where heat loss occurs within the structure, outside losses will go to the environment or the ground mass. Most outside designs incorporate a buried storage tank with the ground acting partially as an insulator to buffer temperature fluctuation. If the water table is too high, this can prevent underground storage and force you to go to aboveground alternatives such as well-insulated sheds attached to the structure.

A storage area which is integrated into an exterior surface of the structure often offers the best alternative. Walls of high thermal mass can act as passive solar collectors as well as performing an active storage and architectural function. The drumwall and the Trombe wall are characteristic of this approach. Figure 38 shows two simple systems in which passive and active elements are integrated. Note that care must be taken to adequately insulate these areas and prevent nighttime heat losses to the outside. In addition, venting and damping louvres are recommended to enhance circulation and ventilation patterns.

Solar Energy for the Northeast

Storage containers may be fashioned from a variety of materials including fiberglass, steel, iron, concrete (poured or block), wood and elastomerics. The material chosen will depend to a great extent on the size of the storage container and the medium. Also the type of system being considered will have an obvious effect on the storage configuration. Three system types will be examined in detail. These include:

        Domestic hot water storage
        Hydronic space heating storage
        Rock bed space heating storage

Details for designing either water or rock storage systems are provided to aid in the formation of a sound decision. It is safe to assume that if you are contemplating the use of a hydronic type solar collector, then a water storage system will be employed, and in the case of an air type collector a rock bed system will be in order.

DOMESTIC HOT WATER STORAGE

In the domestic hot water system, the storage requirements will probably not exceed 120 gallons. In fact, for a family of four an 80-gallon tank will serve quite adequately. The normal estimate for hot water consumption is about 10 gallons/day/person. In other words, that four-person family consumes 40 gallons per day for showers, dishwashing and laundry purposes. A solar hot water system sized to provide 80 gallons of 120°F water will then have two days storage. It is recommended that provision be made for at least two days storage. In this manner, a sunny day may be followed by an overcast day without interrupting the supply of solar hot water.

The solar hot water tank must always be installed in conjunction with a backup water heater. This backup system may be either electric, oil or gas fired. It will insure that during periods of extended cloud cover there will remain an adequate supply of hot water. The backup system is installed in series with the

4: Solar Storage

LOCATING THE SOLAR PREHEAT TANK

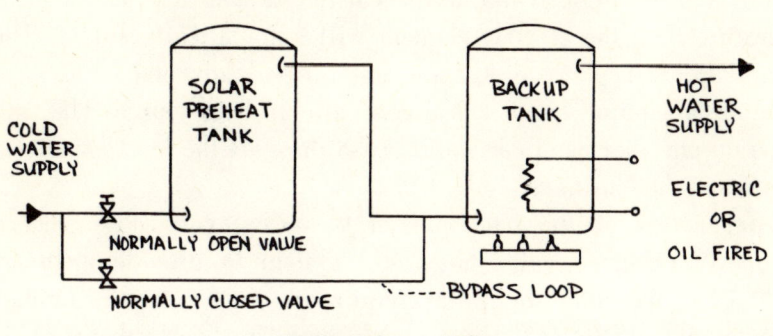

FIGURE 39

solar preheat tank as illustrated in Figure 39. Notice that cold water (50-60°F) from the city main or a well is delivered first to the solar preheater. The water is then warmed through a heat exchanger configuration. As you draw hot tap water in the home the backup tank gradually empties and refills with hot water from the solar tank. Because the backup tank is receiving water already hot, its thermostat will not activate the burner system. Conventional energy is conserved.

It is obviously important that insulation on both tanks be capable of retaining both the solar heat and the conventional heat. It is certainly counterproductive to provide hot water to the backup tank if that tank is going to rapidly cool and activate its burner. It is also helpful if hot water consumption is concentrated in the evening hours. This guarantees that an adequate supply of hot water produced that day is available for use that night before any cooling or tank heat loss has occurred. Admittedly, most bathing is done in the morning, though dishwashing and laundering may be conveniently staggered to evening periods.

The choice of a backup system is limited; it will probably be the existing hot water heater in the home. If you are building a new house, two alternatives are available. You may use a one-

Solar Energy for the Northeast

tank system as shown in Figure 40. In this configuration, an electric heating element is placed directly in the solar tank. If solar energy is not available to heat the tank to the desired temperature then the electric element will automatically cut in. The other option is more costly and has no real advantages; it is to include a separate tank with a conventional heat source. The way current energy costs look, natural gas presents the least expensive results in the Northeast.

Regardless of the backup energy source, you can expect a properly designed solar hot water system to provide virtually 100% of the domestic water requirements during the summer months and up to 75% on a yearly basis in the Northeast. This will translate to between 300 and 500 kilowatt hours of electricity costing $15.00 to $20.00 per month. A savings of $250.00 per year is not unusual at present utility rates. If you are using oil as a backup, then estimating your savings is somewhat more difficult because you will need to determine what percentage of your oil supply is used in providing hot water as compared to space heating. By examining the 5 month period of May thru September you should find the monthly oil requirement for hot water only; multiplying this over a 12 month basis the yearly requirement will result.

SINGLE TANK SYSTEM

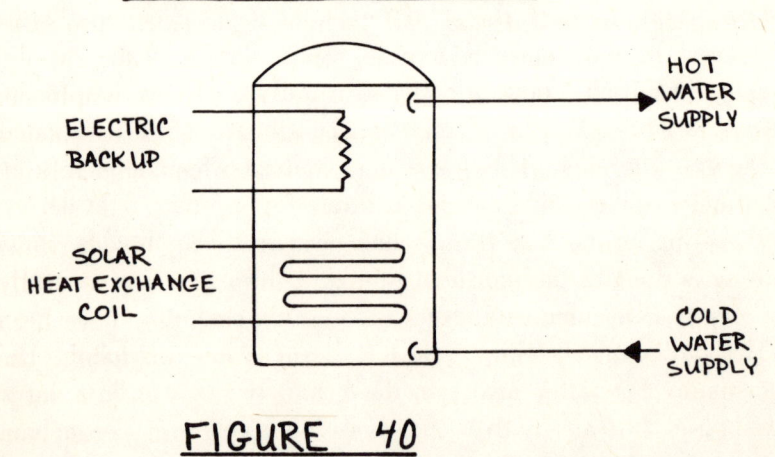

FIGURE 40

4: Solar Storage

Currently, natural gas may be considered a relatively cheap energy suorce. If you are fortunate enough to have a gas fired water heater, then you may postpone your choice of solar hot water for a few years. Inevitably, however, the cost of natural gas will also rise to a point where the solar alternative is cost-effective.

Returning to the operation of solar domestic water systems, it is helpful to understand the process which transfers heat from collectors to storage. This is accomplished through the use of a heat exchanger between the collector circuit and storage. The type of heat exchanger employed will influence the design of the storage tank. Four major configurations are shown in Figure 41 and described below.

Internal coil heat exchangers consist of a coil of copper tubing, with or without aluminum fins, located within the storage tank. Hot fluid coming from the solar collector outlet passes through this coil, losing its heat to the much cooler water in the tank. Constant daytime cycling in this manner warms the tank to a point which is close to the collector outlet temperature. The outlet temperature may be adjusted by controlling the flow rate on the collector array. A slow flow rate will mean higher temperatures whereas a fast flow will result in lower outlet temperatures. A collector system providing 130-140°F outlet fluid is generally the standard.

Drawbacks to the internal coil include replacement problems in the event of corrosion, and contamination of the water supply. To prevent contamination, a double wall coil may be employed, though this reduces the heat exchanger effectiveness. Advantages of the internal coil are its simplicity of design and high heating efficiency.

External counter flow heat exchangers are preferred where the storage capacity is exceptionally large. These units are generally not required on domestic hot water systems, but they have been employed in many cases simply out of force of habit. The advantage is that in heating large volumes of water, a large temperature difference between the solar circuit and storage can

Solar Energy for the Northeast

HEAT EXCHANGER CONFIGURATIONS

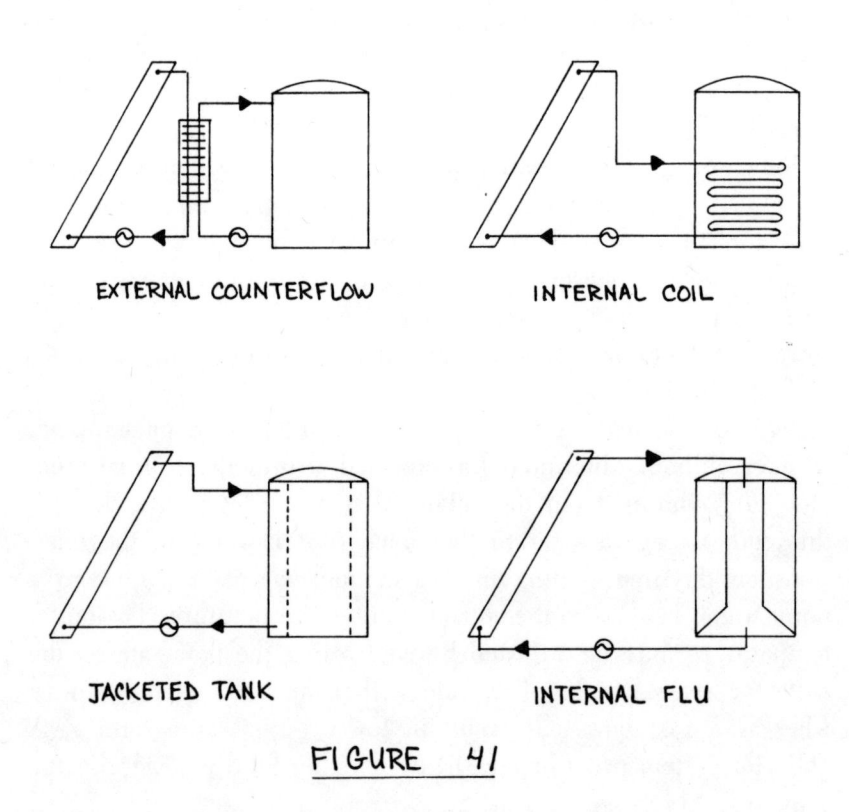

EXTERNAL COUNTERFLOW

INTERNAL COIL

JACKETED TANK

INTERNAL FLU

FIGURE 41

be maintained. This is because the cold water at the bottom of the tank is constantly being pumped through the exchanger, delivering the warmest water to the top of the tank. A simple coil internal to the tank may have a tendency to heat only that portion of the tank in which the coil is located. This is not a problem in small tanks as the force of heat rising due to convection currents will evenly heat the tank. Large tanks with internal coils require mixing devices to aid in convecting heat away from the coil.

The external heat exchanger also requires the addition of another pump in the storage system which will increase costs. The

4: Solar Storage

problem with contamination in the event of a corroded exchanger remains, unless the double wall type is used.

Jacketed tank heat exchangers allow the heated solar fluid to flow through a jacket which surrounds the water supply. In this case, the entire tank wall becomes the heat exchanger surface. The counter flow mode is preserved if the solar circuit moves from top to bottom and the water supply moves in the opposite direction. This, again, encourages the highest possible temperature difference between storage and the heating medium.

Jacketed tanks require special attention to insulation since the heating fluid is located on the outside wall of the tank. A solar circuit at 130°F will tend to lose heat to a 50°F basement as fast as a 50°F water supply, unless some thermal barrier is employed.

Internal flue heat exchangers are somewhat similar to jacketed tanks in that a large surface area is available for heat transfer. A flue which is located inside the water supply will not have as great a tendency to lose heat out the sides of the tank. The nature of the construction involved in these units often represents a higher cost.

In all cases, heat loss from the storage tank must be closely controlled. Attention to the tank insulation is well worthwhile, considering the minimal cost involved. You may easily insure an efficient storage tank by purchasing fiberglass batts and wrapping both the solar preheat tank and the backup tank with extra insulation. While you are doing this, check your backup tank for a temperature regulator. Be sure to reduce the setting on this thermostat to 120°F; many commercial tanks are preset at the factory as high as 180°, which is absolutely unnecessary unless you prefer steam baths. The extra energy that goes into providing the additional 60°F temperature is twice as much as you require. This is also true for hot water systems that work off your oil burner; in most cases, you can lower the setting for domestic hot water independently from that for space heating.

In order to observe the performance of a solar hot water system, you need only locate two thermometers in the collector circuit. A thermometer on the inlet and outlet of the heat exchanger will

Solar Energy for the Northeast

show how much thermal energy is being left in the storage tank. If these two monitors read identically, the pump is not circulating fluid and the system is inoperative. A working system will have an inlet temperature in the range of 80 to 140°F with an outlet temperature of 60 to 120°F. Insist on these two inexpensive items when purchasing from a manufacturer; they are the only reliable means of knowing that the system is working and not just looking pretty.

Solar preheater tanks are available in galvanized steel, glass lined steel, stone lined steel, copper and fiberglass. Standard sizes range from 40 to 120 gallons. It is important to choose a high quality tank at the onset. Replacing a storage tank which may weigh over 500 pounds is not something you will look forward to, so consider locating the tank in an area which is accessible.

HYDRONIC SPACE HEATING STORAGE:

Two schools of thought exist on hydronic storage capacities. One states that since most of the solar energy becomes available during the summer, then it is this heat which must be stored for winter use. The other school counters that summer storage requires tremendously large volumes of water and is therefore impractical; rely instead on winter solar gain to provide daily heat requirements. The result to the homeowner is a choice between 5,000 to 10,000 or 1,000 to 2,000 gallon storage tanks. A 5,000-gallon tank is not small and may require up to 50% of your basement area. With this tank, however, you can potentially store nearly 5 million Btus of summer solar energy. This occurs if you have 5,000 gallons of water at 60°F and raise it to 180°F. (5,000 gallons × 8.3 lbs/gal × 120°F differential = 4,980,000 Btus). Of course, heat losses will occur which reduce this capacity but, by the same token, additional heat gain will take place whenever the sun comes out during the winter. Locating this 668 cubic foot storage mass which weighs 20 tons is difficult but quite possible. Many precautions must be taken to assure no details have been

4: Solar Storage

overlooked which may later jeopardize the safety of the entire structure.

A relatively inexpensive and yet reliable method, if designed properly, is to incorporate the storage tank into the poured concrete foundation of the home. As illustrated in Figure 42, the concrete forms both the foundation and the storage reservoir. Note that special consideration must be given to loading requirements. If the water level is kept low (i.e. 5 ft.) then the 20 tons will be dispersed over a larger area and loading per square foot is reduced. Regardless of this, special consideration must still be given to the foundation floor, support footings, and soil bearing capacity. A competent concrete contractor can handle these requirements safely.

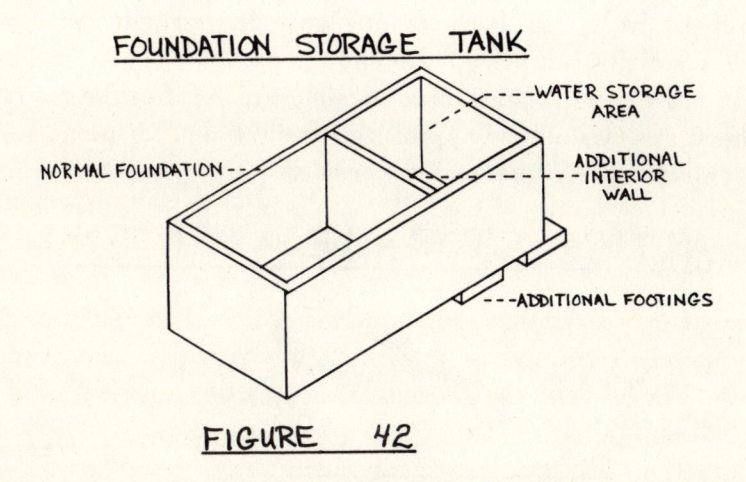

FOUNDATION STORAGE TANK

- WATER STORAGE AREA
- NORMAL FOUNDATION
- ADDITIONAL INTERIOR WALL
- ADDITIONAL FOOTINGS

FIGURE 42

In addition, a suitable liner within the tank must be provided; your local swimming pool contractor is familiar with this item. Many reliable liners or epoxy paints are available to waterproof the tank. Choose carefully to insure that leakage will not result.

The storage water should also include additives to demineralize the water and maintain a neutral pH. This will prevent corrosion and scale buildup on piping and heat exchangers; it will also curtail the growth of algae, mold and mildew. Chlorine may also be added as a precautionary measure.

Solar Energy for the Northeast

Several other methods of storage have been used successfully. These include large fiberglass tanks which may be buried outside. In this case, access ports must be left available in the event it becomes necessary to service heat exchangers below ground. The greatest drawback to a buried fiberglass or steel tank is access for maintenance purposes.

In addition to fiberglass, various precast concrete shells are also available which may be suitable for outdoor installations. Wood stove tanks have been considered for indoor or above ground locations where capacities up to 2,000 gallons are required. Also, concrete block may be arranged with vertical and horizontal reinforcement rods to add strength to a small 1,000 gallon tank.

Work has been done with elastomeric (rubber) pillows that are fitted into crawl spaces or attics for hot water storage. A novel passive system is shown in Figure 43 and indicates how a little creativity can often solve storage problems.

In short, the question of designing and locating large water storage systems can be approached from many perspectives. As long as you have thoroughly considered the potential hazards,

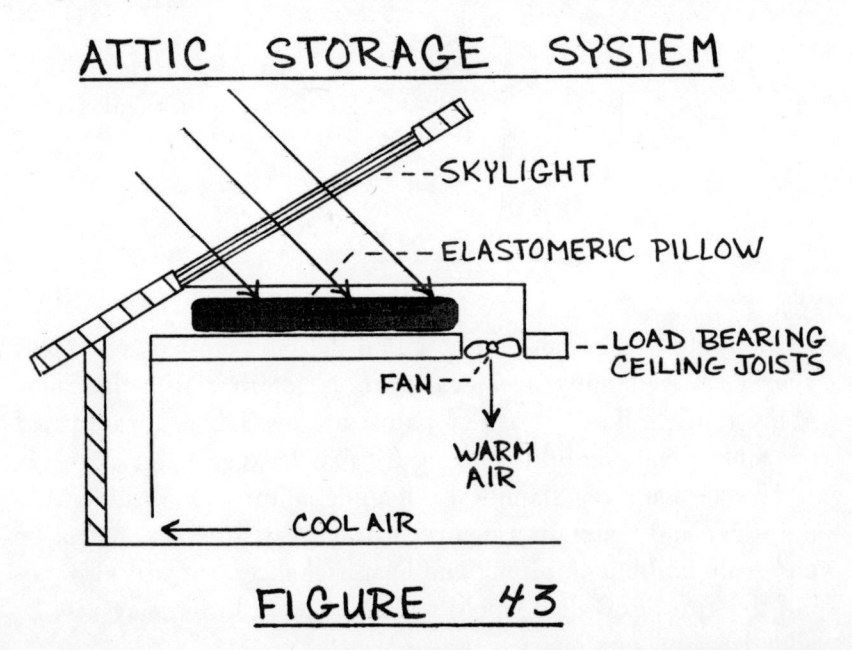

FIGURE 43

4: Solar Storage

VARIOUS TANK CONFIGURATIONS

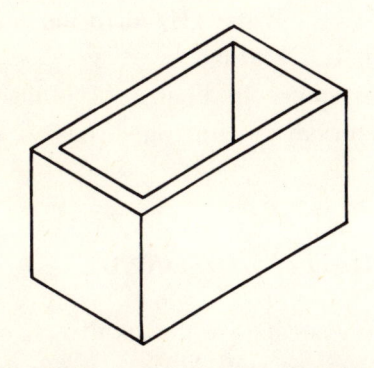

PRECAST CONCRETE

WOODEN STAVE

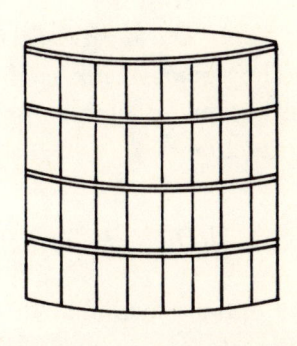

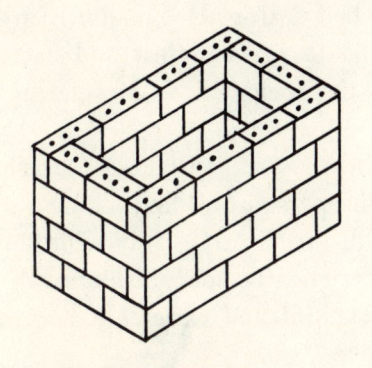

REINFORCED CONCRETE BLOCK

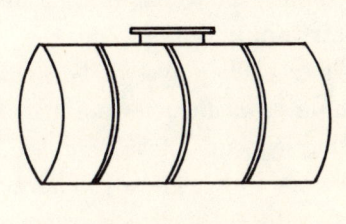

FIBERGLASS

FIGURE 44

Solar Energy for the Northeast

you may rest assured that the storage poses no threat to the structure in which it is located. In review these points include:

Tank loading requirements
Heat losses and humidity
Tank location/water table/frost line
Durability of the tank shell
Prevention of algae and mildew
Maintenance access
Water pH/corrosion and scaling

The various tank configurations shown in Figure 44 indicate the options available for water storage systems as discussed in this section.

ROCK BED SPACE HEATING STORAGE

The options available in rock bed systems are not as diverse as water storage methods. Poured, precast and block concrete or wooden bins are most commonly used. Structural loading becomes a more severe problem as you are now dealing, in some cases, with more than 50 tons of rock. A typical rock bed system is shown in Figure 45. Notice the use of a plenum at the bottom of the bed to facilitate air flow through the bed uniformly. Standard furnace filters may be used on the outlet to prevent dust and bacteria, which may be accumulating in the rock bin, from entering the hot air supply duct.

The rock bed system is certainly less expensive and extremely durable from the viewpoint of the person building his own solar heating system. These beds are also used exclusively with hot air collectors which are also characteristically home-fashioned. Their disadvantages of low storage potential and extremely large volume, however, cannot be overlooked.

Choosing the size of the rock will depend on the configuration of the storage bed. According to the design and dimensions, rocks between 1 and 4 inches in diameter may be used. Often, sources for a particular size of gravel may not be easy to find. Suppliers are accustomed to providing gravel for fill and road construction purposes, but not solar storage bins. Attempts to purchase a uniform rock size may be met with disappointment.

4: Solar Storage

ROCKBED STORAGE BIN

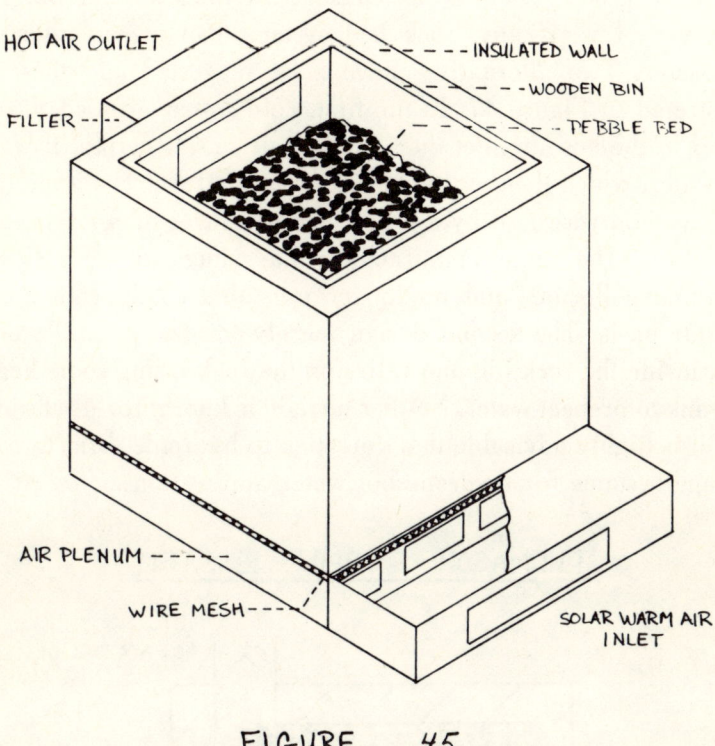

(Figure 45 shows a rockbed storage bin with the following labeled parts: HOT AIR OUTLET, FILTER, INSULATED WALL, WOODEN BIN, PEBBLE BED, AIR PLENUM, WIRE MESH, SOLAR WARM AIR INLET)

FIGURE 45

Considering the volumes required, it is unlikely that you will be able to acquire the necessary amount without going to a special supplier. A rock bin with heat storage capacity similar to a 1,000 gallon water tank will be almost three times as large; this is due to the low specific heat of rock as compared to water.

An innovative combination of water and rock storage has been devised by Dr. Harry Thomason of Solaris Solar Systems. Shown in Figure 46, this system consists of a storage tank which is buried within a rock bed. In this manner the tank slowly loses its heat to the surrounding rocks. As air is circulated through the voids

99

Solar Energy for the Northeast

between the rock it picks up heat and transfers it out to living spaces. The rock becomes a slow heat exchanger in this configuration rather than a storage medium.

The rock bed system is not efficient for domestic hot water; in fact, very few, if any, rock bed systems can provide domestic hot water. Two alternatives have been suggested and these are illustrated in Figure 47. In the first, cold water is piped through a coil in the hot air inlet duct. As hot air moves across, it warms the water, which then moves to a conventional water storage tank. This method was used by a major manufacturer of air type solar collectors. The same manufacturer has since added a line of hydronic collectors and no longer uses this configuration on a regular basis. The second design merely locates a small storage tank inside the rock bin and relies on the rock losing some heat to the tank to preheat water. Neither method is known for its efficiency and it is highly advisable that you stick to hydronic collectors and storage systems for domestic hot water applications.

THOMASON STORAGE SYSTEM

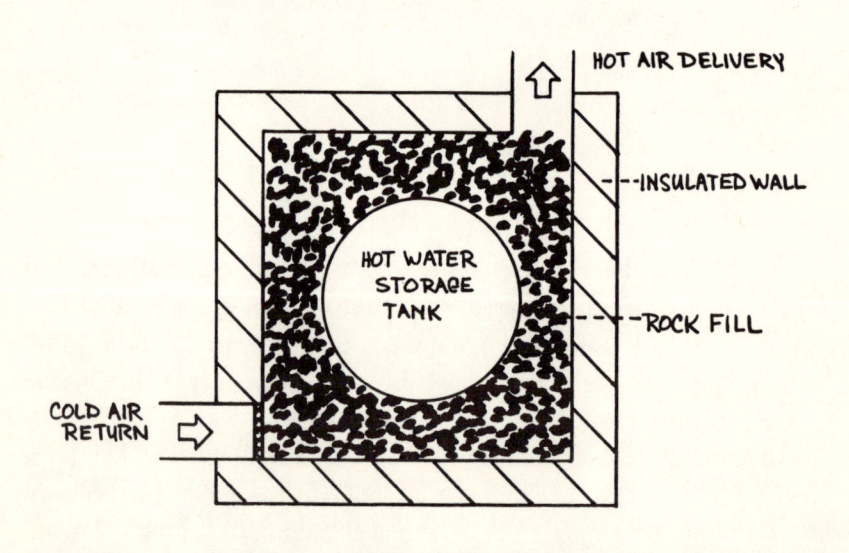

FIGURE 46

4: Solar Storage

HOT WATER SYSTEMS WITH ROCK STORAGE

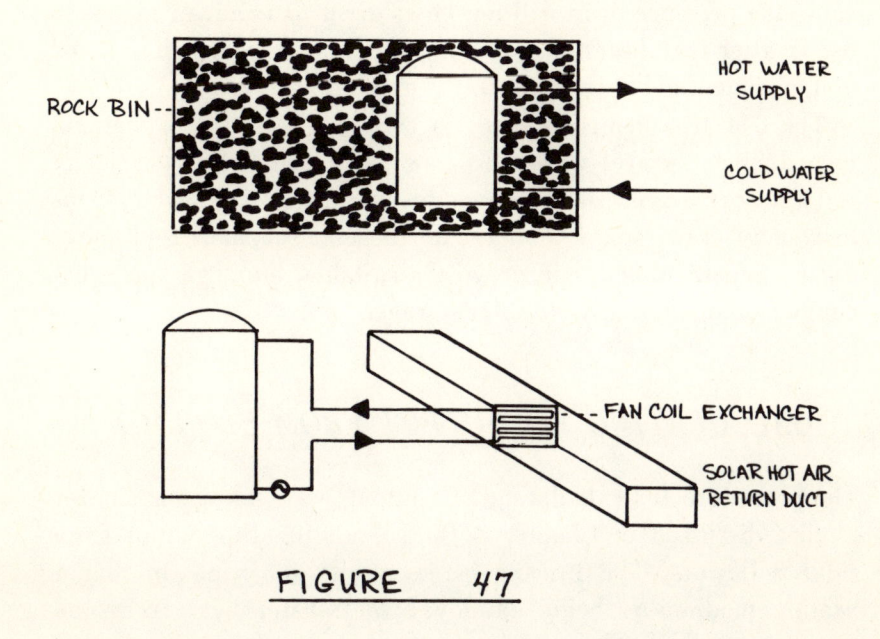

FIGURE 47

Rock bins can be devised which fit inconspicuously into a building's interior. For instance, Dr. George Löf of Denver developed an interesting alternative for rock storage in his Colorado home. Pictured in Figure 48 are vertical cylinders filled with loose solids. These cylinders can be integrated into the building's interior quite unobtrusively. Air may be circulated through or around the uninsulated cylinder to provide space heating.

The shape of the storage bin and the path of air flowing through it will influence the diameter of rock involved. In general, the longer the distance the air must travel through the bed, the larger the rock must be. This is easy to understand if you consider that the air is traveling through the air spaces between the rocks. A fine mixture of gravel will obviously present higher resistance to flow and therefore a high pressure drop between storage inlet and outlet. Figure 49 indicates this effect and the relative sizing of rock diameter to overcome it.

Solar Energy for the Northeast

The velocity of the air also becomes a consideration at this point. At low velocities smaller rock diameters can be used because the pressure drop will not be as great. It is advantageous to use smaller rock because in this manner surface area is increased and the potential for heat transfer is greater.

The overall conclusion then, in rock bed systems, is that 1-2 inch diameter gravel with short flow paths and low flow velocity is to be preferred. Based upon this premise, rock bed systems located in crawl spaces underneath the entire home, if properly ducted and regulated, can provide a suitable, although somewhat cumbersome, alternative to solar storage.

CALCULATING Btu OUTPUT FROM STORAGE

The procedure for calculating Btu output from storage is similar to that discussed in Chapter 3 for calculating Btu output from solar collectors. The process is essentially the same in that a heating medium is being employed to transfer heat from one place to another. The material will again move at a specified

VERTICAL COLUMN STORAGE

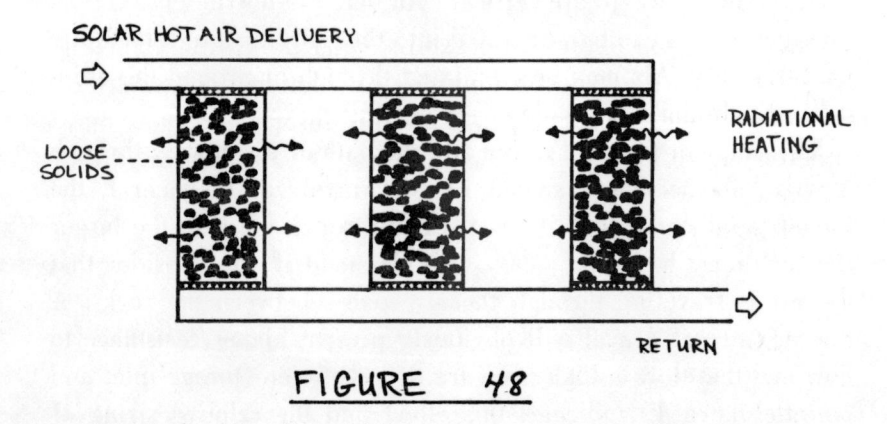

FIGURE 48

4: Solar Storage

AIRFLOW IN ROCKBED SYSTEMS

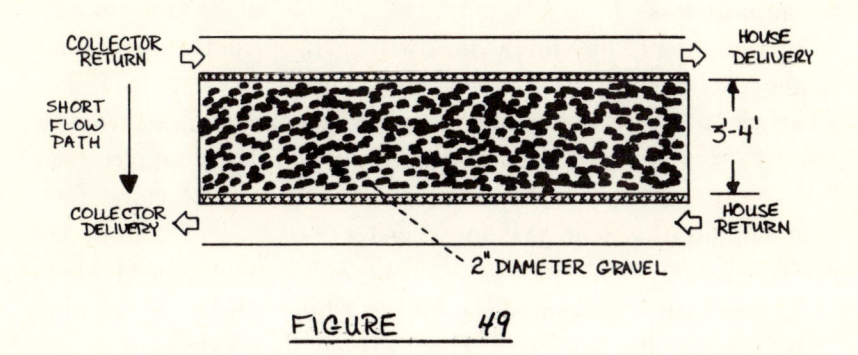

FIGURE 49

mass flow rate with an observable temperature difference between inlet and outlet points. The equation is as follows:

Storage Output = Mass Flow Rate × Specific Heat × Temperature Differential

The storage output is given in Btu/hour and indicates the useful heat gain to the living space.

In hydronic storage systems, the mass flow rate will be in gallons per hour multiplied by 8.3 lbs. per gallon; in air systems the equivalent units are in cubic feet per hour multiplied by .072 lbs. per cubic foot. The specific heat of water is 1.0 Btu/lb/°F and that of air is .24 Btu/lb/°F. The temperature differential remains a function of the temperature outlet minus the temperature inlet in °F. After performing this calculation you can determine the output from storage. With this information you may compare output from storage with output from solar collectors. A more detailed examination of this topic is given in the chapter on solar systems.

CALCULATING HEAT LOSSES FROM STORAGE

The rate at which a storage system loses heat will be a function of the surface area, insulation, storage temperature, and ambient

Solar Energy for the Northeast

temperature. Although you may have little control over the ambient temperature, the remaining three variables can be adjusted to minimize losses.

Surface area of the storage tank will change with the dimensions of the tank. For instance, a rectangular tank 4' × 4' × 8'4" has a surface area of 164.8 square feet, while a tank with the dimensions 4' × 10' × 3'4" has a surface area of 158.4 square feet. Both tanks have a roughly equivalent volume, 132 cubic feet. The configuration with the lower surface area is, of course, the preferred one.

Storage insulation can be chosen according to R-factor. A value of R-20 is usually adequate. The corresponding value for use in heat loss computations is the U-factor. There has been a great deal of confusion over when and how to use these two coefficients. The U-factor is merely the inverse of the R-factor; in other words, $U = 1/R$ or an R-factor of 20 is equal to a U-factor of .05. The R-factor, or resistance to heat transfer, is useful to the consumer because as R increases the quality of the insulation increases. In this manner, it is quite simple to compare insulation materials according to performance. The U-factor, or rate of heat transfer, is useful to architects and engineers in computing heat losses. It is plain to see that the rate of heat transfer is logically the inverse of the resistance to heat transfer.

The temperature at which storage is maintained will influence the temperature differential between storage and ambient. For instance, if the outside air temperature is 30°F and the storage temperature is 150°F, then a 120°F difference exists. However, if the storage is 110°F then only an 80°F difference exists, and the rate of heat loss decreases. The storage high temperature range may be manipulated by adjusting storage volume. A tank holding 1,000 gallons at 150°F contains 830,000 Btus. (assuming the tank's original temperature was 50°F). A tank containing 1,667 gallons at 110°F also holds 830,000 Btus even though the overall tank temperature is 40°F lower. Other factors must be taken into consideration, however, before jumping into a larger tank. The cost, loading requirements and space consumed by the larger

4: Solar Storage

storage area may be more important than the relatively small difference in rate of heat loss. Or you may simply utilize additional insulation to offset heat loss at higher storage temperatures. In either case, care must be taken to insure that the storage temperature is not too low to extract the heat effectively. A 5,000 gallon tank at 70°F may contain 830,000 Btus, but transferring this energy to a 60°F living space will be quite difficult unless a mechanical heat pump is employed. The cost of these units is often restrictive.

To calculate heat losses from a storage tank, let's first look at an uninsulated tank of precast concrete with 3 inch walls. The U-factor for 3 inches of concrete is .68 (R-factor is 1.47). The tank is a cube 5 feet on each side and therefore has 150 square feet of surface area. Multiplying .68 by 150 we arrive at a heat loss of 102 Btu/hr/°F. If the tank is 150°F and the outside air is 30°F then a 120°F difference exists. Multiplying this 120°F difference by a heat loss of 102 Btu/hr/°F we arrive at an overall heat loss of 12,240 Btu/hr or 293,760 Btu/day.

The same tank with an R-20 insulation material (U-factor is .05) will have 150 square feet times .05 or a heat loss of 7.5 Btu/hr/°F. Again, if the difference between storage and ambient is 120°F then 120°F times 7.5 Btu/hr/°F yields an overall heat loss of 900 Btu/hr or 21,600 Btu/day. With the R-20 insulation, almost 93% of the heat which would have been lost is contained. Considering the relatively low cost of insulation as compared to other components in the system, the resulting gain in thermal efficiency is significant.

If you recall that a Btu is the amount of energy required to raise one pound of water by 1°F, you can rapidly compute that with sufficient insulation, tank losses should not be greater than 3°F per day. When designing storage systems every effort should be made to insure that losses do not exceed this limit on a daily basis.

Many insulating materials are available which meet this requirement, including fiberglass, urethane foam, isocyanurate and glass fiberboard. Vermiculite and similar loose fill insulations are

Solar Energy for the Northeast

not recommended as they have a tendency to both settle and compact. Polystyrene is inadequate unless the tank is expected to remain below 150°F.

Key points to remember when choosing a tank insulation include resistance to moisture, fire and compacting, as well as R-value or insulating quality.

5: SOLAR DISTRIBUTION

Once having collected solar energy and transferred it to a suitable storage area, the remaining function becomes distributing this energy effectively for end uses. A tank full of hot water or a rock bed containing warm pebbles is of little use unless the heat contained there can be transferred efficiently, minimizing losses along the way.

The low grade energy contained in a solar storage area represents a difficult transfer problem for several reasons. Often the storage temperature may be only slightly higher than the end use temperature. For example, heating a room to 70°F is much easier with 180°F water than with 100°F water. This is because heat will transfer quickly and relatively effortlessly when a large temperature difference exists. With a small differential the conditions must be closely regulated in order to insure maximum and efficient transfer.

In addition, there is little room to accommodate heat losses in transfer if the storage supply is very close to the heating requirement. Extreme losses will mean that storage is depleted before the room heating requirement is fulfilled. This in turn will cause the heating system to cross over to the backup oil burner or electric heater, a case which is certainly not preferred.

Solar Energy for the Northeast

METHODS OF HEAT TRANSFER

There are three means by which heat is transferred. It is necessary to have a clear comprehension of each of these in order to understand the overall processes of solar collection, storage and distribution. The methods of heat transfer have not been discussed previously in either of the chapters on collection or storage although they have certainly been involved. In solar distribution, however, heat transfer comes into the forefront. In all cases, it is useful to remember that there is one common denominator, a golden rule of heat transfer, so to speak, which although obvious is often not considered in sufficient detail. This rule has two parts.

1. Heat will always try to move from areas of high temperature to areas of low temperature.
2. The greater the difference between the high temperature area and the low temperature area, the harder the heat will try to move.

The three methods of heat transfer are conduction, convection and radiation.

In *conduction* heat is passed from molecule to molecule until a uniform temperature exists throughout the object. For example, a metal spoon immersed in boiling water will soon be as hot at the handle as at the submersed end. Heat is being conducted out of the boiling water and up the spoon. Some materials conduct heat better than others. Wood is a poor conductor and therefore a wooden spoon will not heat up when immersed in boiling water. Metals such as copper, aluminum and iron represent the best conductors.

Heat transfer by *convection* is accomplished when a fluid is passed over a heat source causing it to cool and release its energy to the fluid. In the case of convection, the fluid may be either liquid or gas (water or air). The most common example is the cooling of hot soup by blowing across it.

5: Solar Distribution

FORCED AND NATURAL CONVECTION

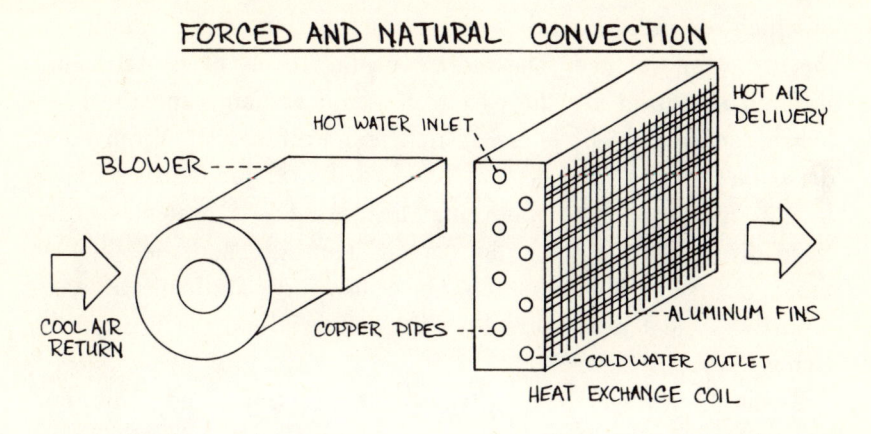

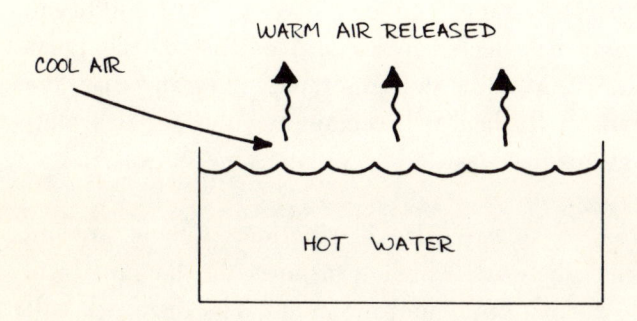

FIGURE 50

Convection may be forced or it may take place naturally. Figure 50 illustrates both these principles. Notice that natural convection, the force of heat rising, is a common form of passive solar heating because no pumps or fans are required. The analagous forced convection system employs a pump which circulates hot water through a coil and a fan which blows across the coil to release and transfer the heat. This fan/coil arrangement is the most common form of active solar heating systems.

Radiation, the third form of heat transfer, is the result of electromagnetic thermal waves which are transmitted from the surface of the heat source. Any hot object will radiate its heat. The rate at which the radiation takes place is a function of the emission of the radiating surface. The most familiar forms of heat transfer through radiation are the cast iron steam radiator and the baseboard radiator. In both cases, high temperatures are required to drive these units. The typical baseboard radiator consists of ¾ inch copper pipe with aluminum fins. As 180°F water is circulated through the system, the copper heats up, and the adjacent fins heat up due to conduction. Once they have attained the maximum possible temperature they begin to radiate the heat to the living space.

These three forms of heat transfer, convection, conduction, and radiation, may be arranged in a variety of ways depending on the type and temperature range of the storage media. In addition, the distribution may be effected by natural or forced techniques as a result of passive or active design strategies. In any case, the media used to transfer the heat will remain as either air or water, just as the storage media consisted of either stone or water.

Air Distribution Systems—Air distribution systems are installed by locating ductwork which transports the hot air from a central point to rooms throughout the home. The ductwork will consist of both supply and return ducts which are carefully sized to provide adequate flow and insure uniform heat distribution. Louvers or vents may be manually operated to prevent the heating of spaces which are not in use. For instance, bedroom vents may be closed during the day while living or family room vents are opened. At night the situation can be reversed.

The ductwork will probably be part of a wood or oil burning furnace system in which the combustion is used to heat air which is then circulated by a fan through the ducts.

In addition to the conventional heating system, a solar rock bed or water storage system may be integrated into the duct work. A fan is used to draw heat out of the rock bed and introduce it into the distribution ducts. If the rock bed temperature is not sufficient to meet heating requirements, a two stage thermostat may be used to cross over to the conventional heating system.

5: Solar Distribution

The rock bed may also be installed independently from the conventional system. As shown in Figure 51 a rock bin located in a crawl space requires only two openings to serve as inlet and outlet ducts. At the outlet side a fan is used to draw out hot air in the stone pile and deliver it to the living space.

There is a broad variety of configurations in which the rock bed may be employed. Whether located in basement, crawl space or living space, the basic form of heat transfer remains convection. In all cases, cool room air is being drawn across the warm rock, cooling the rock, and transporting the heat to living spaces.

Air distribution systems used in conjunction with solar water storage systems are by far the most widely employed form of solar space heating. As described previously, the hot water is pumped out of the solar storage tank and circulated through a coil consisting of copper tube and aluminum fins. The coil is located in the return side of the main furnace duct adjacent to the existing furnace fan. A two stage thermostat may again be used. In the first stage, when the home requires heat, both the pump and the fan will turn on but not the oil burner. In this manner, solar heated water is pumped through the coil and the fan draws cool air across the coil to remove the heat and distribute it to living spaces.

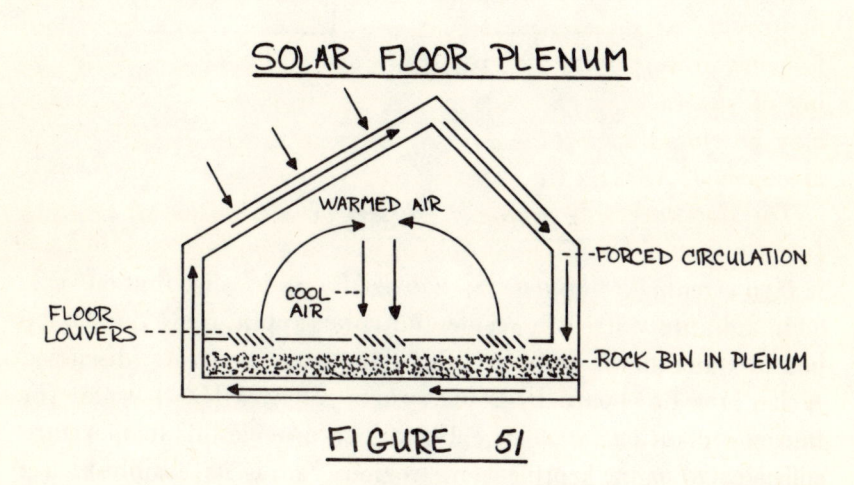

SOLAR FLOOR PLENUM

FIGURE 51

Solar Energy for the Northeast

In the event that the solar water is not hot enough, the second stage of the thermostat will automatically kick in the oil burner to provide makeup heat. A summary of this system is depicted in Figure 52.

The fan/coil system is widely used because useful heat can be extracted from water at temperatures as low as 90°F. In solar storage tanks, the average temperature may often drop to this extreme. In order to achieve efficient heat transfer across the coil, special attention must be given to coil sizing, fluid flow rate and air flow rate.

HOT AIR DISTRIBUTION SYSTEM

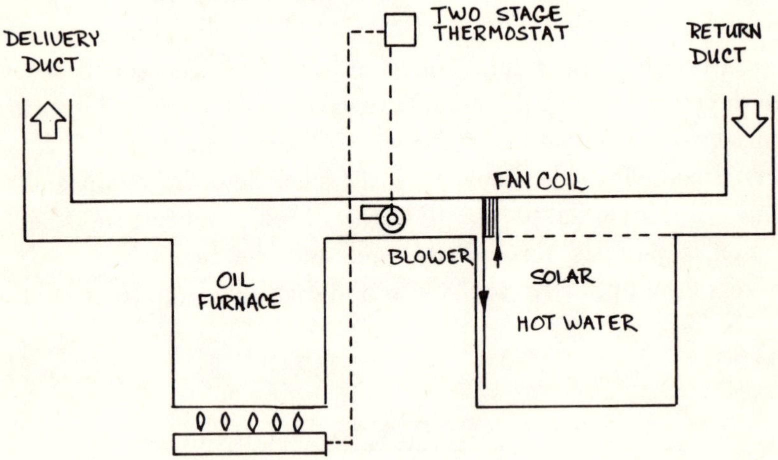

FIGURE 52

Hydronic Distribution Systems—Hydronic distribution systems are rarely used in residential solar applications due to the high temperatures required for their operation. As discussed earlier, the baseboard hydronic system requires 180°F water for efficient operation. No solar collector can provide this temperature consistently in the Northeastern region. Nor is it feasible to use solar to augment an existing baseboard system although this is

5: Solar Distribution

by far the most widespread request solar designers have been confronted with.

The only hydronic system which is practical is the radiant floor heating technique. This strategy employs a combination of heat transfer by conduction and radiation principles. In a radiant floor system, hot water from solar storage is pumped through a grid of pipes which are embedded in a poured concrete floor. The result is that heat is conducted from the pipes to the concrete mass and thereby warms the concrete. The concrete, in turn, having reached sufficient temperature, will begin to radiate the heat to the living spaces.

If solar storage temperature is high enough to cause the concrete to radiate heat, a wonderfully efficient and simple distribution system results. It is this author's opinion, however, that the concrete will warm as the embedded piping grid cools but it will never attain a high enough temperature to effect any substantial radiational heating. The overall result is a room in which the floor is quite warm and comfortable but the surrounding ambient air temperature may be only 50 to 60°F. This allows the solar radiant floor heating system to be quite effective for garages, basements and work spaces but only marginally effective in living spaces. Its primary advantage is that a warm floor will lead to a comfortable environment provided you keep your feet on the ground. It is the only system that relies almost totally on conductance as its means of heat transfer. You can see that solar "radiant" floors are actually a misnomer and that solar "conductance" floors are closer to the actual conditions.

The solar conductance floor is illustrated in Figure 53. Note that a relatively low water temperature is required to bring a concrete floor to 70°F. This means that out of all solar distribution systems the conductive floor requires the least energy input and will therefore provide the most consistent and sustained output during the winter season.

Heat Pumps—A solar storage area may contain several hundred thousand Btus of stored energy. When it comes time to distribute this energy for end uses, we often find that up to 20%

113

Solar Energy for the Northeast

HYDRONIC DISTRIBUTION SYSTEM

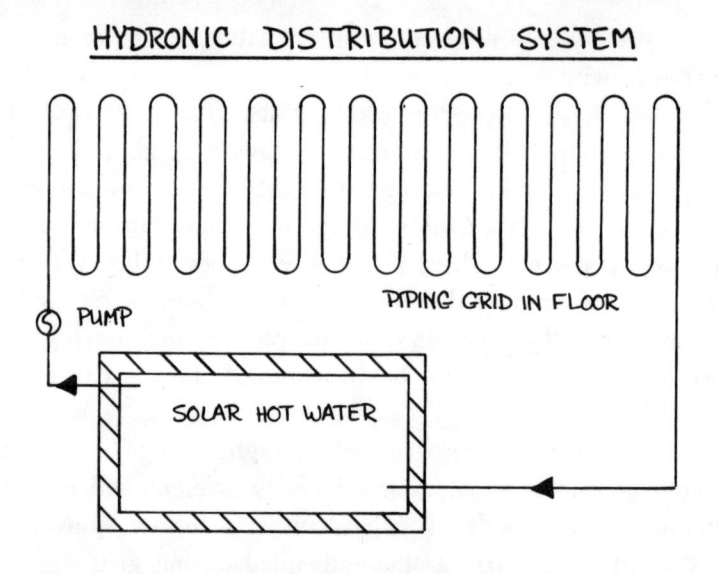

FIGURE 53

of this energy is not usable. The problem with unusable energy results from the fact that the end use temperature may be quite close to the storage temperature. With only this small difference between the warm area and the cool area, the tendency for heat to flow is relatively low. For instance, consider this example: a 500-gallon storage tank is at 80°F; if the cold water that entered this tank came in at 60°F, then it was solar energy which raised that water temperature by 20 to 80°F. More specifically, it was 83,000 Btus of solar energy, if you recall that it takes 1 Btu to raise 1 lb. of water 1°F. That 83,000 Btus is equivalent to 1 gallon of fuel oil. (Assuming that #2 Fuel oil contains 132,800 Btu/gal and burns at 62.5% efficiency.)

Since the storage tank is at 80°F, there is no practical means to transfer the energy efficiently. A fan coil requires 100°F water to be effective, a conductive floor 90°F and a baseboard radiator 180°F. Every day, this solar system will have 83,000 Btus which are unusable due to the small temperature differential. If a heating system operates 100 days a year, that is the equivalent of 100 gallons of fuel oil. And that's with a small 500-gallon tank!

It is easy to see that the low grade energy which remains unusable in a storage area is truly a waste. Also, another factor comes into play. If you remember that a collector system operates more efficiently at lower temperatures, you will certainly see the advantage in lowering that tank temperature to 60°F before beginning a new day of solar collection. The rewards are twofold—net energy gain is increased as the system works more efficiently at lower temperatures, and usable energy is increased as you apply low grade heat to your requirements.

The heat pump is designed to accomplish these goals. A device used for both heating and cooling, the heat pump represents an evasive concept which has evolved with the help of space age technology. The transfer of heat is accomplished by circulating a refrigerant which changes between its liquid and gaseous state as a result of a series of pumps, compressors, condensers and valves. The latent heat involved in the phase change is drawn out of the solar storage tank when the refrigerant evaporates, and is dissipated to the living space when the refrigerant re-condenses.

A liquid-to-air heat pump can extract low grade energy (down to 45°F), concentrate it, and deliver it at higher temperatures which are usable. There is, of course, the inevitable disadvantage. Heat pumps require electricity to drive the pump and compressor. The output energy is a combination of heat which results during compression of the refrigerant and recondensing of the vapor. The efficiency of a heat pump is given in terms of its co-efficient of performance (C.O.P.). This value is the ratio of heat obtained to heat equivalent to the electrical energy used to operate the system. A C.O.P. of 3 to 1 is considered quite good.

In general, the use of heat pumps has been limited to date in favor of lower cost distribution systems. Although the distribution end of a solar system is often the least costly item, every effort is made to incorporate it into the conventional system and capitalize on its dual purpose. Any investment in improving the distribution system is, of course, an improvement in the overall performance. The decision as to just how far to go to attain additional performance is a function of economic resources available.

Simple modifications such as insulating heating ducts are low cost and certainly more sensible in the long run than the addition of exotic equipment like heat pumps which have debatable economic justification.

6: SOLAR SYSTEMS

Having thoroughly examined the elements of solar collection, storage and distribution, we can now integrate these elements in the overall design of a useful solar energy system. This integration is fairly obvious: consider that air collectors will most probably be installed in conjunction with rock bed storage and forced air distribution systems. Likewise, in most cases, a hydronic collector will include hydronic storage and fan/coil forced air distribution.

The variety of design strategies and applied strategies implies that useful solar systems are by no means cut and dried, definitive systems. Variations in system planning arise as a result of several important and highly user-oriented criteria. These criteria include:

Site restrictions and limitations: the amount of solar energy available at the site will ultimately influence the extent to which useful energy may be derived from the system. The direct influence on design strategy will be the choice between a full scale system or a small supplemental system.

Structural design and aesthetics: the complete utilization of both passive and active solar elements most often requires a somewhat radical departure from traditional architectural aesthetics. Roof slopes and southern building facades demand a new examination in the light of energy consumption by the structure. The overall heat loss and gain must be thoroughly considered and modifications made in order to insure that energy is conserved.

Degree of user involvement in system operation: we have the technological capability to design completely automatic solar heating systems. This capability results from the wide range of electronic monitors and sophisticated control systems available on the market. The cost of this convenience increases with the system's intricacy. It is also possible, and at quite a lower cost, to design manually operated systems which require the user to be aware of the environment in which he lives.

Technical capability of the designer: many simple active systems and most passive systems may be designed and installed by the homeowner provided that some ground work has been covered to provide a complete understanding of the principles involved in efficient energy transfer. Full scale active systems, on the other hand, require mechanical engineering services by a competent contractor. The size of pumps, fans, ducts, piping, storage and collector areas cannot be a result of "rough estimates" if an efficient system is to result. Many attempts have been made to "play by ear" the construction of a solar energy system. In many cases, these attempts are marginally effective at best. The homeowner who has some degree of technical expertise can expand his knowledge through the review of existing literature and systems and then determine if he has the qualifications to design a reliable system.

System capabilities: as we have seen previously, solar energy may be used to heat space, heat water, and provide year round growing conditions for plants. Any or all of these applications can be integrated in a system. The degree of precision with which the system operates is also a question. If you wish to maintain an environment between 68 and 70°F then there is little room for the error which may result from lag periods in solar storage or distribution. On the other hand, if an environment above 60°F is adequate, then the control system may be simplified. A manual

6: Solar Systems

switch may be employed which requires you to actively turn on the heat when you are cold. I have witnessed several homes in which this philosophy was employed and the resulting energy awareness among family members was truly inspiring.

Financial capabilities: little can be done to change this influence on the choice of a solar system. Full scale active heating systems are expensive; a range of $12,000 to $20,000 is not unusual. Passive systems and small scale water heating systems, however, are often quite affordable and fall in a range from between a few hundred dollars for a passive solar greenhouse to $3,000 for a fully automatic and well engineered hot water system. Federal and state governments are the best sources for aid in this area. The legislation for energy tax credits has the ability to allow all people to take advantage of the full potential of solar power. This legislation is changing rapidly and a complete investigation of the alternatives should be made prior to the design phase of any solar system.

With a knowledge of the above elements and their influence on system design you can arrive at a reliable estimate of just how extensive a solar commitment you are prepared to make.

Often an analogy is drawn between the cost of solar equipment and the cost of electronic calculators. The analogy goes something like this: "look how inexpensive hand-held calculators have become after only a few years, certainly solar technology will do the same." This premise is incorrect for several reasons. The manufacture of solar collectors is extremely labor intensive and requires large amounts of raw materials. Specifically those raw materials include copper, aluminum and glass, all of which are subject to spiraling inflationary costs as a result of mining, refining, and transportation expenses. Mass production is not the answer to reducing costs as many manufacturers are already engaged in large scale production of solar collector units. And finally, there will be no massive technological breakthroughs in collector design. The processes involved in the transfer of low grade heat have been known and understood for years. A few improvements have been made in the last few years as a result of the new and overwhelming demand for energy. These improvements have been accomplished primarily through the aid of NASA (National Aeronautics and Space Administration) as an offshoot of appropriate technology developed for the space program. The improvements, although individually minor, have collectively resulted in highly efficient and reliable solar collectors. Such items as low reflectance glass, low emittance coatings and lightweight structural components may be credited to space program research.

The systems which utilize solar energy are, from the engineering perspective, old systems. Systems which transfer heat using either a hydronic or air medium are not unusual. Only the solar collectors themselves may be considered new, and their integration into existing frameworks has been both rapid and successful as evidenced by the growing proportion of solar supplemented structures.

In many cases a focusing of attention on the basic elements is required. Control of heat loss and gain, efficient re-distribution of heat, and re-evaluation of end use requirements are the primary objectives in efficient systems design.

The following pages will demonstrate systems which have been constructed and installed and are presently producing usable energy for residential applications. Four major categories are examined which include passive air, passive hydronic, active air and active hydronic. In some cases, the systems discussed have been developed primarily through the research of a specific designer, in which case that designer and his current affiliations are noted. Additional information as well as detailed designs may often be available by contacting the design firm directly.

PASSIVE AIR SYSTEMS

By far the most basic application of solar energy, the passive air system, has the beauty of simplicity; no pumps or fans are required, costs are low and installation is relatively simple. Although

6: Solar Systems

the usable energy derived may be low, its contribution to the overall heating requirement of a home is significant. Passive air systems, for the above reasons, should be a mandatory requirement on all new structures.

A window is a passive solar collector. Sunlight passes through the glass and transmits to interior spaces where it is absorbed by walls, floors and objects in its path. The window, if it is not properly designed, might even allow more heat loss than heat gain. The most obvious improvements include:

1. Location of the largest area of window space on the south wall with minimum window space on north walls.
2. Use of triple or double glazed windows which incorporate more than one pane of glass with air spaces in between to increase the insulating effect and minimize heat losses.
3. The use of silicone or high grade caulking around the perimeter of the window frame to seal cracks and reduce heat losses due to cold air infiltration.

THERMAL CURTAIN

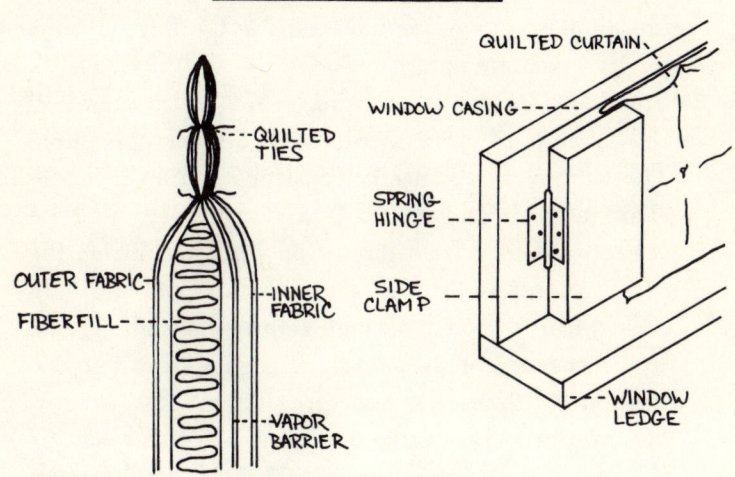

SOURCE: SOLAR AGE MAGAZINE AUG. 79

FIGURE 54

In addition to the above, there are two not so obvious modifications which may be further introduced to increase the efficiency of a passive solar window. These include:

1. The use of a thermal curtain which can be drawn over the window tightly at night, thereby almost totally eliminating heat loss. The curtain material may vary depending on materials available and aesthetic preference. Naturally, a material with high insulating quality is preferred. A simple quilted pattern which can be firmly attached to the window perimeter is often used. Standard curtains and roller shades are not acceptable as these are loosely hung and ineffective in controlling infiltration around the edges. A typical thermal curtain system is shown in Figure 54.

    More sophisticated systems have been employed in which the curtain is mechanically raised and lowered in response to a thermostat. Also, a device known as the Beadwall (Zomeworks Corp.) has been developed which automatically fills the space between windowpanes with styrofoam pellets during the night and removes them during the day with the aid of a vacuum pump.

2. The addition of a solar window box which increases the amount of area available for solar collection. The window box is easily constructed from inexpensive materials and results in a simple solar air type collector. This device takes cool air from inside the home, passes it through the collector with the help of natural convection, and returns hot air to the living space. For greater flow, a small fan may be introduced to aid in the circulation. The window box works only during the day while the sun is shining and has no storage area. If the home is sufficiently warmed due to passive solar gain through windows, then the window box may introduce un-

6: Solar Systems

SOLAR WINDOW BOX

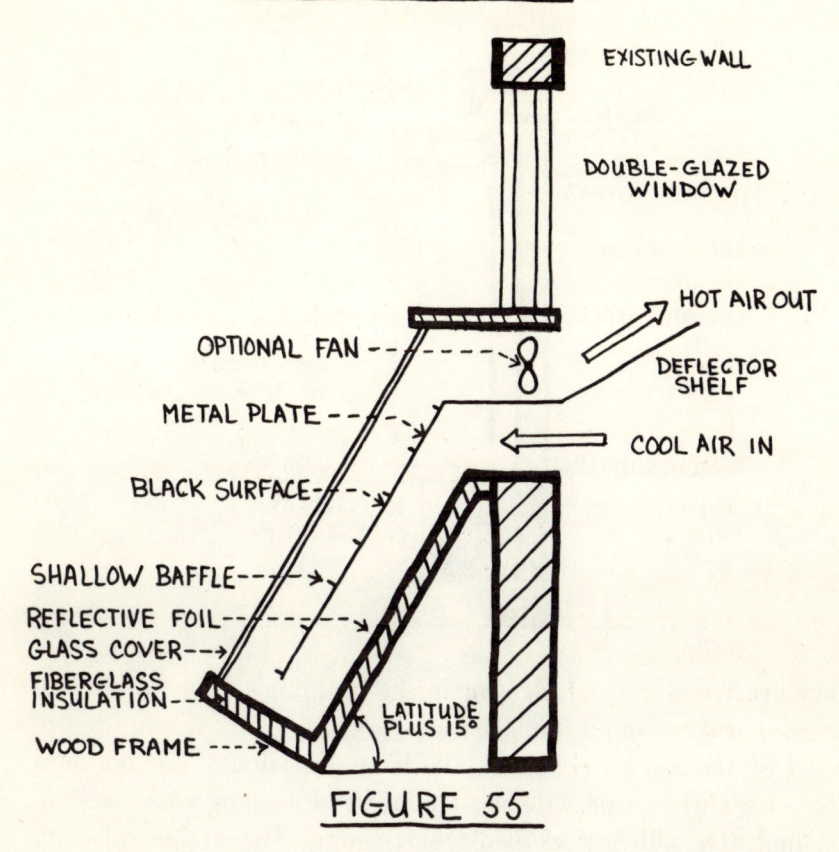

FIGURE 55

Labels: EXISTING WALL, DOUBLE-GLAZED WINDOW, HOT AIR OUT, DEFLECTOR SHELF, COOL AIR IN, OPTIONAL FAN, METAL PLATE, BLACK SURFACE, SHALLOW BAFFLE, REFLECTIVE FOIL, GLASS COVER, FIBERGLASS INSULATION, WOOD FRAME, LATITUDE PLUS 15°

needed heat during the daytime and a damper will be required to prevent overheating. Figure 55 illustrates a simple window box system.

The Trombe wall is an extremely basic device which stores heat in a large concrete or masonry mass. As we discussed earlier, if you can increase the mass of a structure, its ability to resist temperature fluctuations increases. This is the goal of the Trombe wall, to absorb heat during the sunny day and release that heat at night after the sun goes down. The wall will typically be large and massive and must be artfully incorporated into the interior decor. With sufficient planning, the wall may become an

Solar Energy for the Northeast

TROMBE WALL

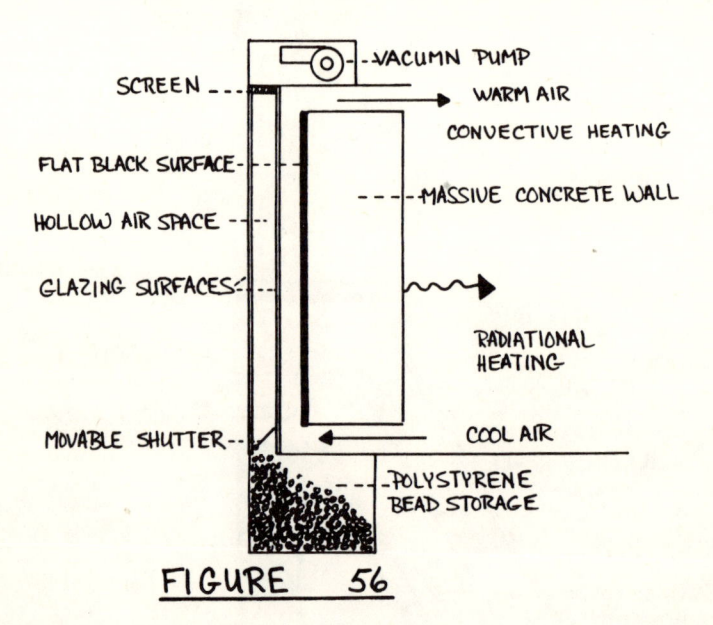

```
                    ┌──○──┐ ── VACUUM PUMP
   SCREEN ───────┤      │
                 │      │ ──→ WARM AIR
                 │      │      CONVECTIVE HEATING
FLAT BLACK SURFACE─┤    │
                 │      │ ──── MASSIVE CONCRETE WALL
HOLLOW AIR SPACE ─┤     │
                 │      │   ～～→
GLAZING SURFACES ─┤     │      RADIATIONAL
                 │      │      HEATING
                 │      │
MOVABLE SHUTTER ─┤      │ ←──  COOL AIR
                 └──────┘── POLYSTYRENE
                            BEAD STORAGE
```

FIGURE 56

unobtrusive structural element in the building's framework. Figure 56 shows a simple Trombe wall.

A passive hot air collector (similar to the window box but on a larger scale) can provide substantial home heating when used in conjunction with a rock bed storage area. The design difficulty arises when you consider that heat rises due to convection and that the storage area must be above the collectors. In addition, the heated space must be above the storage. All this is necessary to accomplish natural circulation without the aid of a fan. The use of a fan transforms this system from passive to active, and introduces new costs associated with electrical wiring, controls and ductwork. A convective system can be achieved with adequate design and planning as shown in Figure 57.

Solar greenhouses are little more than spaces which heat up when the sun is out because they are enclosed by glass or a similar transparent material. As shortwave radiation passes through the glass it is absorbed by the floor, walls or objects

6: *Solar Systems*

PASSIVE SOLAR HEATING

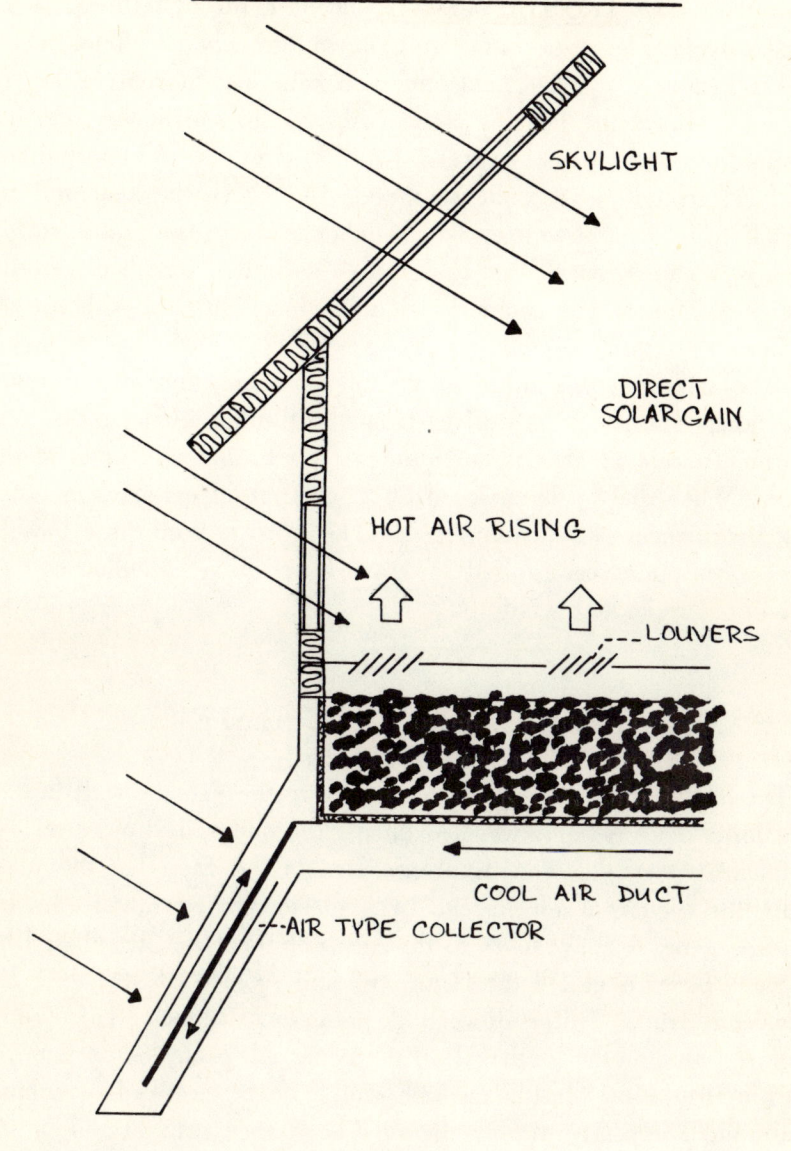

SKYLIGHT

DIRECT SOLAR GAIN

HOT AIR RISING

LOUVERS

COOL AIR DUCT

AIR TYPE COLLECTOR

FIGURE 57

125

within. When the energy is re-radiated from these objects, it is in the longwave form and will not pass through the glass as effectively. The result is the well-known "greenhouse effect." Once this little glass house heats up, you can open a vent or louvre which allows the heat to come rushing into the home. It's that simple. If you place several 55 gallon drums of water, painted black, inside the greenhouse these will absorb the heat and re-radiate it at night to keep the greenhouse warm for plant growth. In either case, recall that those glass walls will lose heat rapidly at night unless you include a double glazed surface with a tight thermal curtain.

Solariums are similar to greenhouses but they sound more elegant. The solarium derives from the Roman atrium concept in which a central area is surrounded on all sides and open at the top. When the top is enclosed by a transparent material in order to retain heat, a solarium results. The word is used quite loosely now and may be applied to anything from a greenhouse to a laundry room with windows.

PASSIVE HYDRONIC SYSTEMS

The passive hydronic system is similar to the Trombe wall in that a large mass is being warmed during the day and allowed to lose its heat at night. The difference lies in the fact that the mass in this case is water instead of a masonry material. There are three general approaches to this method. The first, developed by Steve Baer, is called the Drumwall and consists of stacks of 55-gallon drums, water-filled, and painted flat black. The drums are located on the south wall of the home and protected by a moveable insulation platform which can be hand cranked or motor-driven. The platform seals the outside surface at night and causes the drums to radiate their heat inside to the living spaces.

A roughly similar system developed by Harold Hay and termed Skytherm consists of a pond located on the roof of the home. The same situation is repeated; at night an insulating barrier is moved

6: Solar Systems

over the roof pond causing it to radiate its heat downward into living spaces. Both the Drumwall and Skytherm are illustrated in Figure 58.

Similar to the Drumwall is a concept utilizing water-columns within the living space. Large diameter corrugated piping has been used to store heat for later re-radiation. In all of these systems, the key is locating the mass in an area which will receive daytime sunlight and still be aesthetically acceptable. In each case, the thermal barrier must be employed to prevent heat from radiating back outside.

SKYTHERM BY HAROLD HAY

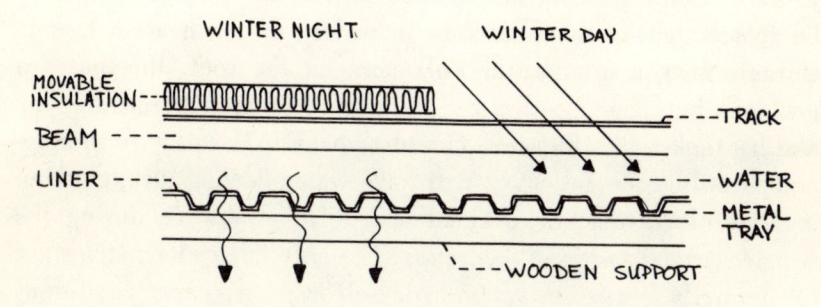

DRUMWALL BY ZOMEWORKS CORP.

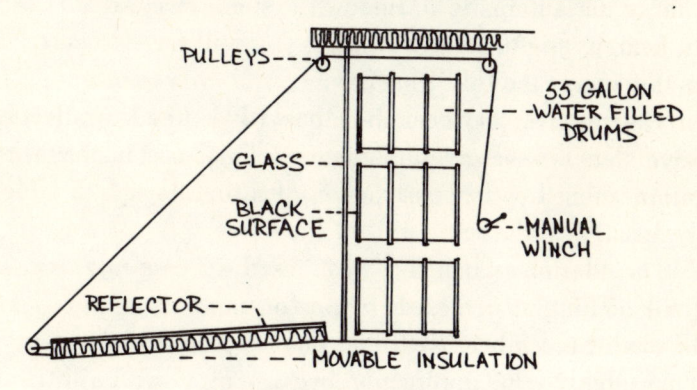

FIGURE 58

An additional advantage of the Drumwall, roof pond or water-column is that the system is reversible. That is to say, during summer months the curtain may be drawn to prevent the water mass from heating up. In this case, heat will be drawn out of the interior of the home with a resulting cooling tendency. At night, when the curtain is again opened, the water mass will cool to the outside environment. If sufficient mass is employed, the temperature fluctuation in the home will be minimized.

A much more widely used strategy employing passive hydronic concepts is the thermo-siphoning hot water system. Instead of heating space as the end use, heating hot water for domestic use is the goal. The thermo-siphon operates on the principle of natural convection. As the solar collector heats up, the warmed water will rise out of the top of the collector and be displaced by cold water at the bottom. The inevitable hitch is that the storage tank must be located above the collectors in order for the heat to rise to storage. If you are putting collectors on the roof, this poses a problem but if the collectors are ground mounted, locating the storage tank is much easier. One drawback to the passive thermo-siphon is that the collectors will have water flowing through them and therefore must be drained and left inoperative during the winter season to prevent freezeups. The only other alternatives are to manually drain the system yourself when freezing conditions threaten or to install automatic draindown valves. Keep in mind that in an automatic draindown system, if a valve sticks or fails to actuate and the water freezes, you will have to go out and pick up the pieces the following day.

The passive thermo-siphon has widespread application in the south but is severely limited to seasonal use in the Northeast. A simple plan for the construction of a thermo-siphon is depicted in Figure 59.

The thermo-siphon has been used successfully for swimming pool heating systems where performance is required from May through September. Redwood pools which are above ground lend themselves to this application because often with a little additional excavation the collectors can be located on a slope below the pool in order to effect the natural convection process.

6: Solar Systems

THERMOSYPHON HOT WATER SYSTEM

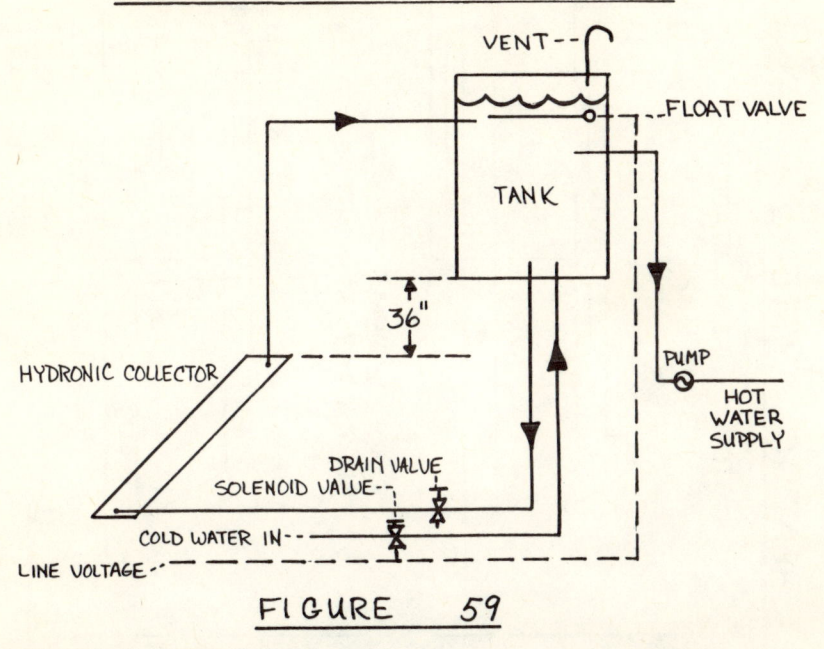

FIGURE 59

ACTIVE AIR SYSTEMS

Solar air heating systems in all cases will use a rockbed storage system. The location of the rock bed is often in either a basement or crawl space, although vertical columns within the living space have also been used successfully.

Hot air is transferred from collectors to storage by ductwork which includes a fan to force circulation. The fan may be sized to produce 3-5 c.f.m. (cubic feet per minute) for every square foot of collector. Conventional furnace filters may also be installed to insure the removal of dust and particles.

Figure 60 indicates a typical solar forced air heating system. Note that the duct system is shared by both the solar and conventional backup system. A two stage thermostat is used to activate the duct fan alone initially. If after several minutes of operation the solar storage area does not provide the required heat, the

Solar Energy for the Northeast

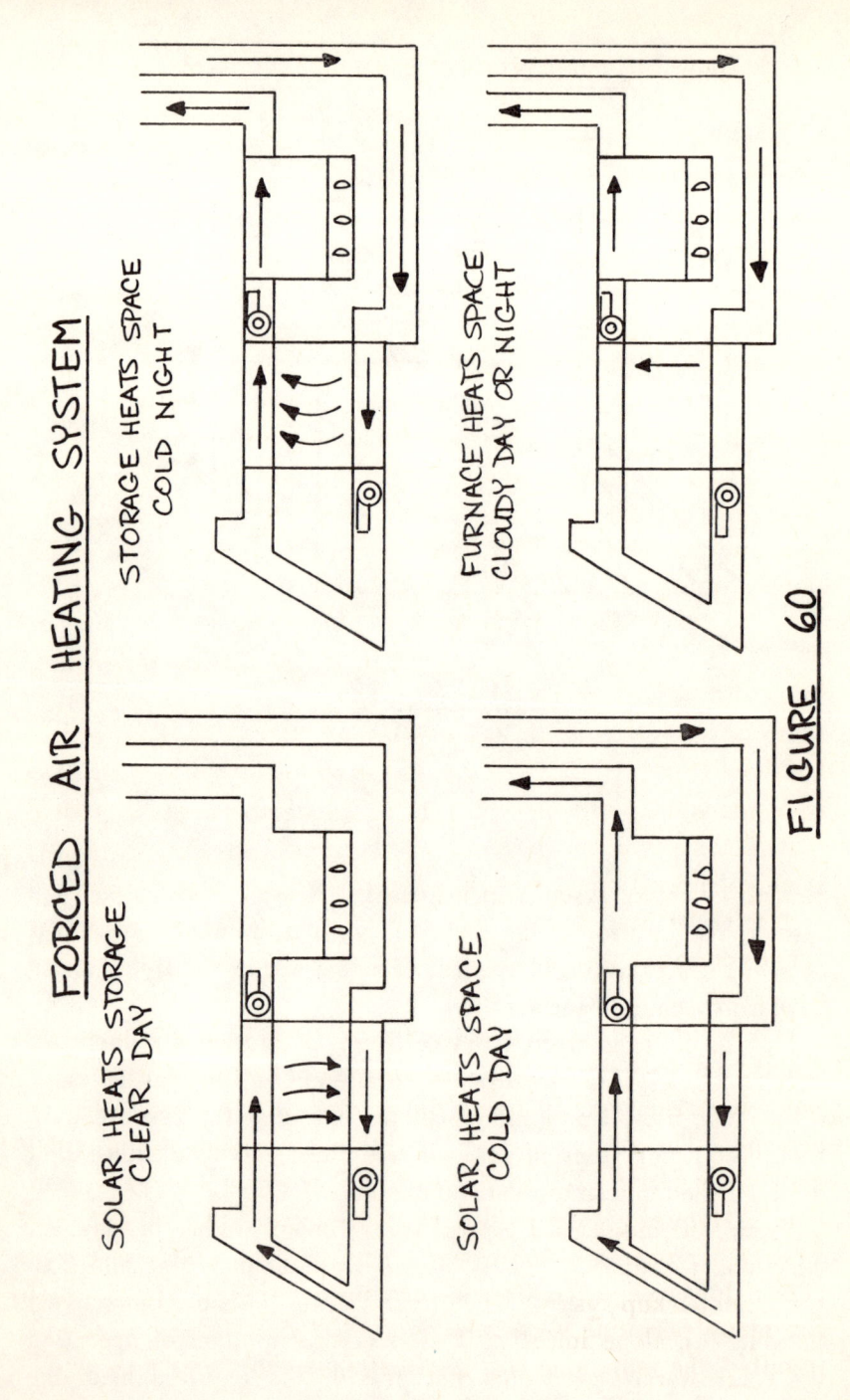

FORCED AIR HEATING SYSTEM

SOLAR HEATS STORAGE
CLEAR DAY

SOLAR HEATS SPACE
COLD DAY

STORAGE HEATS SPACE
COLD NIGHT

FURNACE HEATS SPACE
CLOUDY DAY OR NIGHT

FIGURE 60

130

6: Solar Systems

second stage of the thermostat automatically turns on the oil burner.

Active air systems rely on the daily availability of sunshine, unlike a hydronic system which may store energy in a water tank for extended periods. The rock bed storage capability does not exceed more than a week or two unless unusually large storage bins are employed. This drawback weighs heavily on the successful operation of air systems in the Northeast where the duration of cloud cover may extend to greater than 3-5 days.

The air system nonetheless remains the most inexpensive and easily installed active solar alternative available. Its popularity has grown dramatically as home builders have begun to fabricate their own equipment on-site without the need of a manufacturer to supply pre-made collectors and expensive water storage tanks.

ACTIVE HYDRONIC SYSTEMS

More work has been completed in the development of active hydronic solar systems than in any other solar application. The hydronic system when designed and installed correctly has proven to be both reliable and efficient as a means of transferring energy from collectors to end uses. It is also the most popular form of system to be used for the purpose of generating hot water for simple domestic uses including dishwashing, clothes washing and showering.

Hot water is a common requirement which is often taken for granted and in many instances the fuel used to generate this 140°F water is overlooked as it lies in the shadow of larger home heating demands. The surprising fact is that domestic hot water is the most cost-effectve and widespread use of solar energy at this time. A typical system costing the homeowner up to $3,000 can qualify for nearly a 40% rebate when it comes time to figure personal income taxes. This brings the investment within reach for many potential users who find that after having a system installed, the remaining cost is often recovered within four to six years.

Solar Energy for the Northeast

Hot water systems are available in two general varieties. The first type, known as the *"draindown" system,* allows the water to flow through the solar collectors where it is heated and then back to storage. The approach is simple and easy to attain as there are no heat exchangers required. However, the danger of having water flow where it will be exposed to freezing temperatures is apparent. This restriction forces you to include some provision in the system design which will allow all exterior areas to be drained automatically in the event of low temperatures. The draining process in itself is not difficult as there are many motorized valves on the market which can accommodate this function. The question of reliability and foolproof operation becomes of paramount concern, as the failure of only one valve or sensor will result in water remaining in the system where it is very likely to freeze and rupture collectors or components irreparably. The ensuing cost of repair quickly eliminates any previous savings which may have been realized through the design or purchase of a draindown hot water system.

The draindown systems are quite successful in southern climes where freezing potential is minimal. However the overall recommendation for the Northeast is to avoid draindown configurations unless a very specific and detailed warranty is obtained from both the manufacturer and, more importantly, the system installer. Instructions on maintenance and servicing of valves and controls should also be included at the time of purchase.

If you choose to design and install your own system, the responsibility for freeze protection is your own and must therefore be thoroughly evaluated. The draindown configuration is highly efficient from the heat transfer standpoint because no exchangers are necessary and the specific heat of the fluid, namely water, is 1.0. By virtue of this fact you must also consider the effect of very high temperatures in this system. If circulation is stopped and the water is allowed to stagnate in the collectors, it will rapidly develop high temperatures and pressures which can result in equally disastrous blow-outs.

A reliable and well tried design for draindown systems is

6: Solar Systems

DRAINDOWN HOT WATER SYSTEM

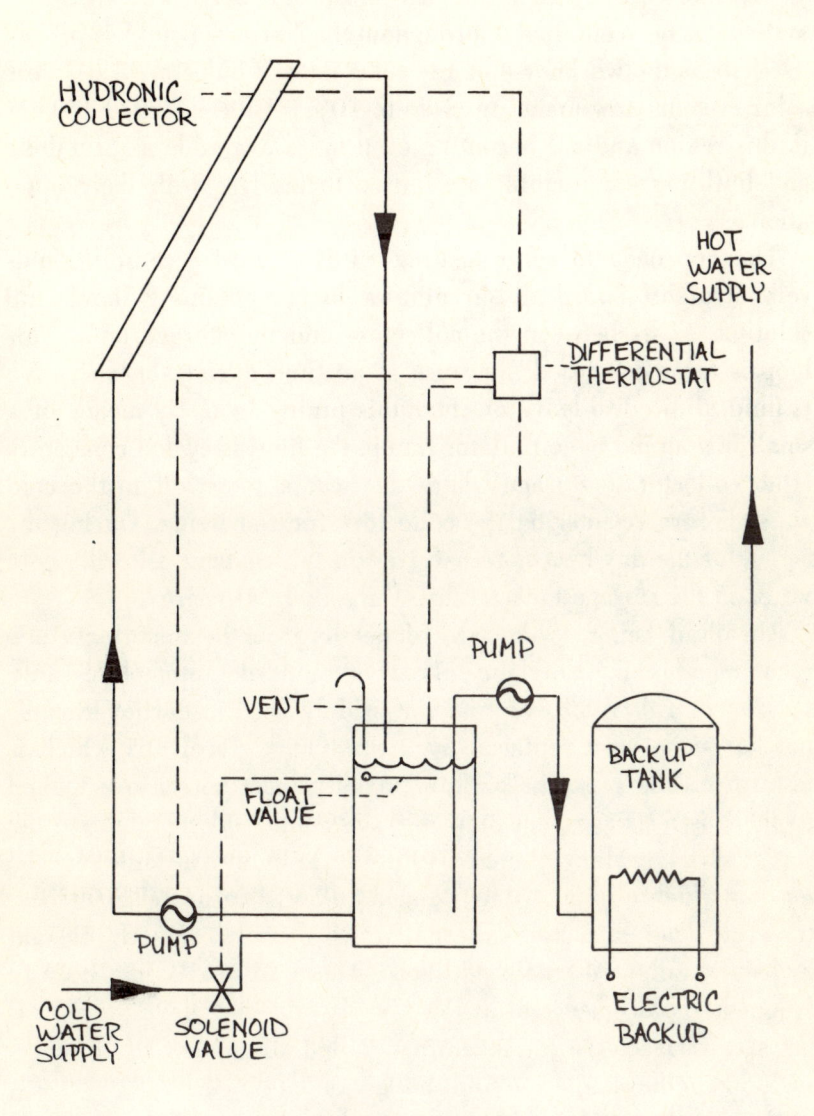

FIGURE 61

133

Solar Energy for the Northeast

depicted in Figure 61. Given that the breakdown of components is ruled out, this system will perform satisfactorily and produce generous amounts of hot water.

The more widely used and less hazardous *antifreeze hot water system* has been employed throughout the Northeast and has proven itself to be the workhorse of the solar energy industry. Antifreeze solar systems now make up close to 90% of the installed systems in this region and are becoming even more available as plumbing and building contractors familiarize themselves with their operation.

The approach to solar heating with an antifreeze solution is very straightforward. A circulating loop, containing the liquid solution, exists between the collectors and the storage tank. This loop is termed "closed" because at no time during the operation is fluid allowed to leave or enter this piping loop. By means of a small low amperage circulator pump, the fluid is cycled constantly from collector to storage where the heat is given off to the cold water before re-entering the collectors for re-heating. During the course of the day heat is transferred in this manner allowing cold water in the storage tank to absorb the thermal energy.

The fluid being used varies depending on the manufacturer's recommendation. Ethylene glycol (standard automobile antifreeze) in a 40-60% solution was widely used in earlier systems but has since been replaced by a propylene glycol mix which is less toxic. The concern for fluid toxicity must not be overlooked as there is a danger that fluid may leak into and contaminate the water supply. Many non-toxic fluids are now available on the market which eliminate this danger and still insure efficient heat transfer. One attractive alternative is the use of light weight hydraulic oils with colored dyes. These oils will not require frequent replacement as do the glycols and they have good heat transfer characteristics. A more detailed discussion of fluids is included in the chapter on solar collection.

Regardless of the fluid being used, the system configuration will remain the same. Figure 62 indicates a typical solar antifreeze system.

6: Solar Systems

ANTIFREEZE HOT WATER SYSTEM

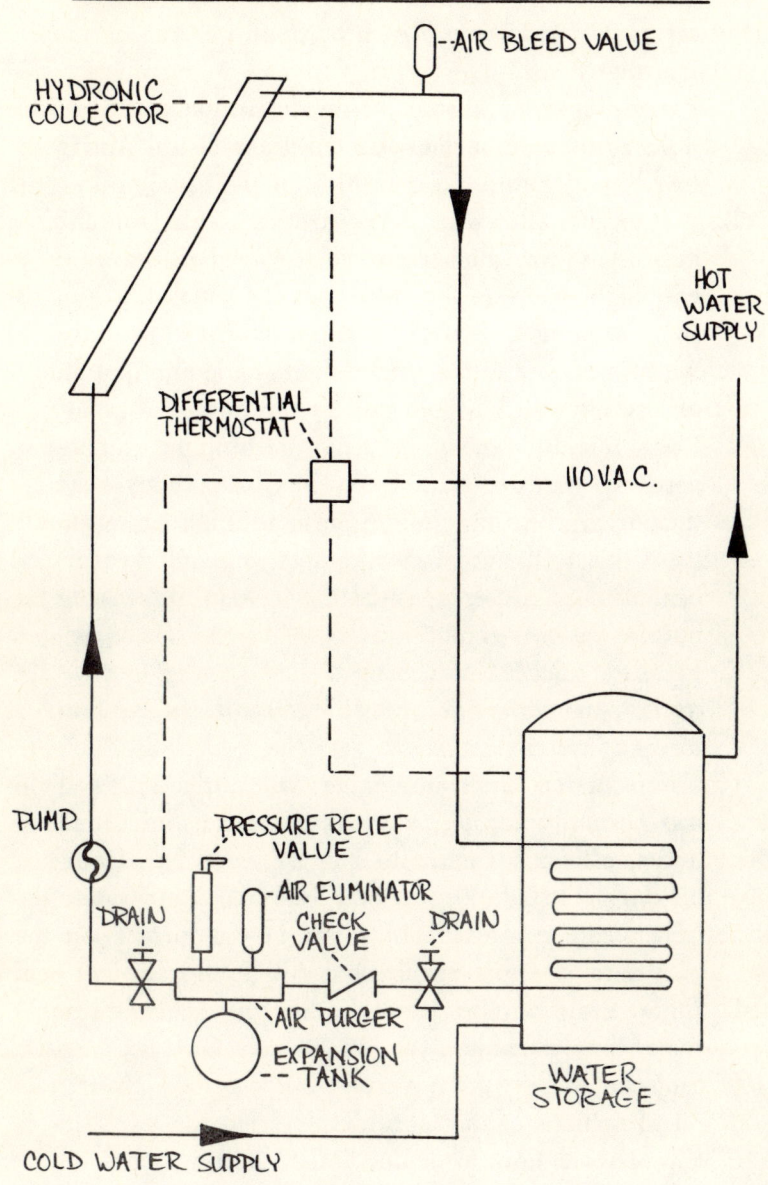

FIGURE 62

135

Solar Energy for the Northeast

You will note the following components which are found in almost all systems presently on the market. A properly engineered solar water heater will contain each of these items to accomplish a specific purpose.

1. Expansion tank: allows for the expansion of heated fluid as its volume increases within a closed piping loop.
2. Air purger: allows for the separation of air from the fluid. Any air which remains dissolved in the fluid will both decrease the heat transfer capability and shorten the useful lifetime of the fluid.
3. Air eliminator: allows for the automatic elimination of any air which is collected by the purger.
4. Pressure relief valve: prevents the buildup of dangerously high pressures in the loop which may exist due to overheating. In the event of high pressure, the valve will automatically open and allow overheated fluid to escape until the loop returns to normal operating conditions.
5. Pressure gauge: indicates pressure on the loop in p.s.i. (pounds per square inch) and allows for monitoring of system status.
6. Thermometers: indicate temperature of the fluid and are normally located at heat exchanger inlet and outlet, allowing for monitoring of system performance.
7. Check valve: prevents backflow in the loop which may arise due to natural convective cycling.
8. Boiler drain valves: allow for the filling and draining of the closed loop.
9. Stop and waste (gate) valves: allow for the closing off of sections of the loop for servicing.
10. Circulator pump: provides forced flow in the loop to cycle fluid between collector and storage. Solar pumps are generally of high efficiency, low amperage design to minimize the electrical cost of operation.

6: Solar Systems

The above listed items should be present along with the collectors and storage tank in any system which you intend to purchase. In addition to these plumbing fixtures, an automatic electronic controller will be included. The controller is known as a differential thermostat and functions by monitoring two inputs and forming a decision to either turn the pump on or off depending on the conditions. The inputs are provided by a sensor located in the collector and another located at the tank. When the collector sensor is higher than the tank sensor, the controller activates the pump and continues its operation until the temperature at the tank sensor exceeds the collector temperature. These controls are long life solid state circuitry which demand no further attention once properly installed.

The use and effectiveness of solar hot water systems, as described above, have been extensively tested and proven in the Northeast. The results of this implementation have led to their rapid success and proliferation in households throughout the region. There is no question as to the most economical use of solar energy: it is, unequivocally, domestic hot water.

There are several alternative uses for solar hot water which have also proven to be economically justifiable. A well known option is the *solar heating of swimming pools.* This is an ideal application of solar power because the requirement exists during the same time of year that the maximum amount of solar energy is available. In the Northeast, the swimming season is generally considered to be June, July and August, and even during these summer months the water temperature in a home pool may drop to 70°F. An effective solar system will not only stabilize the water temperature above 80°F, but also extend the period of use to include May and September. For indoor pools, solar energy can significantly defray the expenses involved in heating the pool on a year-round basis.

The simplicity of a pool heating system results from the fact that the system is not being used during freezing periods. This reduces the complexity of the system tenfold. Water can be circulated through the collectors directly without the use of anti-

Solar Energy for the Northeast

freeze solutions and heat exchangers. Since water has a specific heat of 1.0 it is also extremely effective in transferring the energy. In addition, the storage area is also the end use and the requirement for a distribution system is eliminated.

In general, the use of solar energy for swimming pool heating is a natural; the costs are low and the benefits high. You can fashion your own system inexpensively or purchase a low efficiency commercial solar collector and install the system yourself. Figure 63 indicates the simple circulation system required for

POOL HEATING SYSTEM

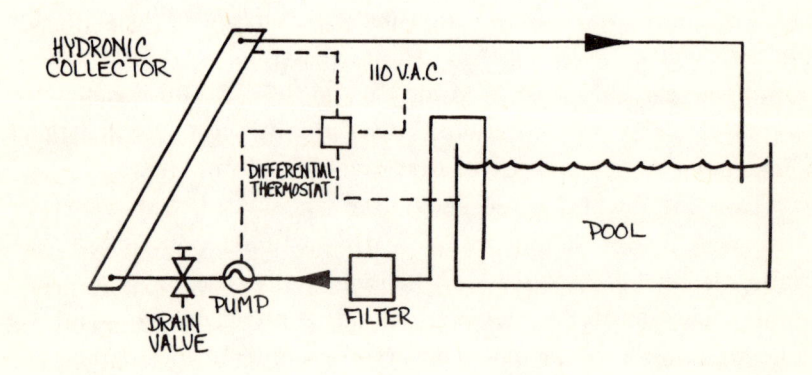

FIGURE 63

pool heating. Little deviation from this basic approach will be necessary. You may circumvent the purchase of commercial collectors by constructing your own from ½ inch copper pipe and corrugated sheet metal. Follow the basic approaches discussed in the previous chapter on solar collection to complete the collectors. Also, recall that you will have to drain the collectors and piping to prevent freezing during the winter.

6: *Solar Systems*

Many so called "pool blankets" are available from pool contractors and supply stores. These blankets are quite effective in preventing evaporative heat loss from the pool and will usually recover their entire cost within two years. The use of a pool blanket is a deceptively simple addition which can save hundreds of dollars in pool heating costs over the life of the system.

The remaining use of solar hot water is space heating by either fan-coil forced air systems or conductive floor systems. The fan-coil system is most prevalent not only in the Northeast but throughout the country. In this system, a coil of aluminum finned copper tubes is placed inside the cold air return duct on your conventional oil burning furnace. Solar heated water is drawn from a storage tank and pumped through the coil. The furnace fan draws air across the hot coil and heats the air in the process. The hot air is then moved into the living space by the existing hot air ducts. Figure 64 summarizes this solar heating strategy.

Each phase of operation in this configuration has previously been reviewed in the chapters on collection, storage and distribution systems. This graphic summary, however, ties together each of the processes and indicates that space heating with solar energy is not a simple task. The interfacing of three independent systems into one integrated, efficient system is a job for the experienced heating, ventilating, and air conditioning (HVAC) engineer.

A somewhat simpler form of solar heating is the conductive floor. As discussed earlier, the conductive floor is an offshoot of the conventional radiant floor heating system. In the familiar radiant system hot water (160-180°F) is generated by an oil fired boiler and circulated through a piping grid which is embedded in the floor. As the water cools the floor heats up and begins to radiate that heat to the living space. In the conductive floor this process stops one step sooner. Solar heated water (80-100°F) is now pumped through the pipe grid and warms the floor. Unfortunately, it is not likely that the floor will reach a high enough temperature to begin radiating that heat but it will warm considerably and provide a comfortable environment.

Solar Energy for the Northeast

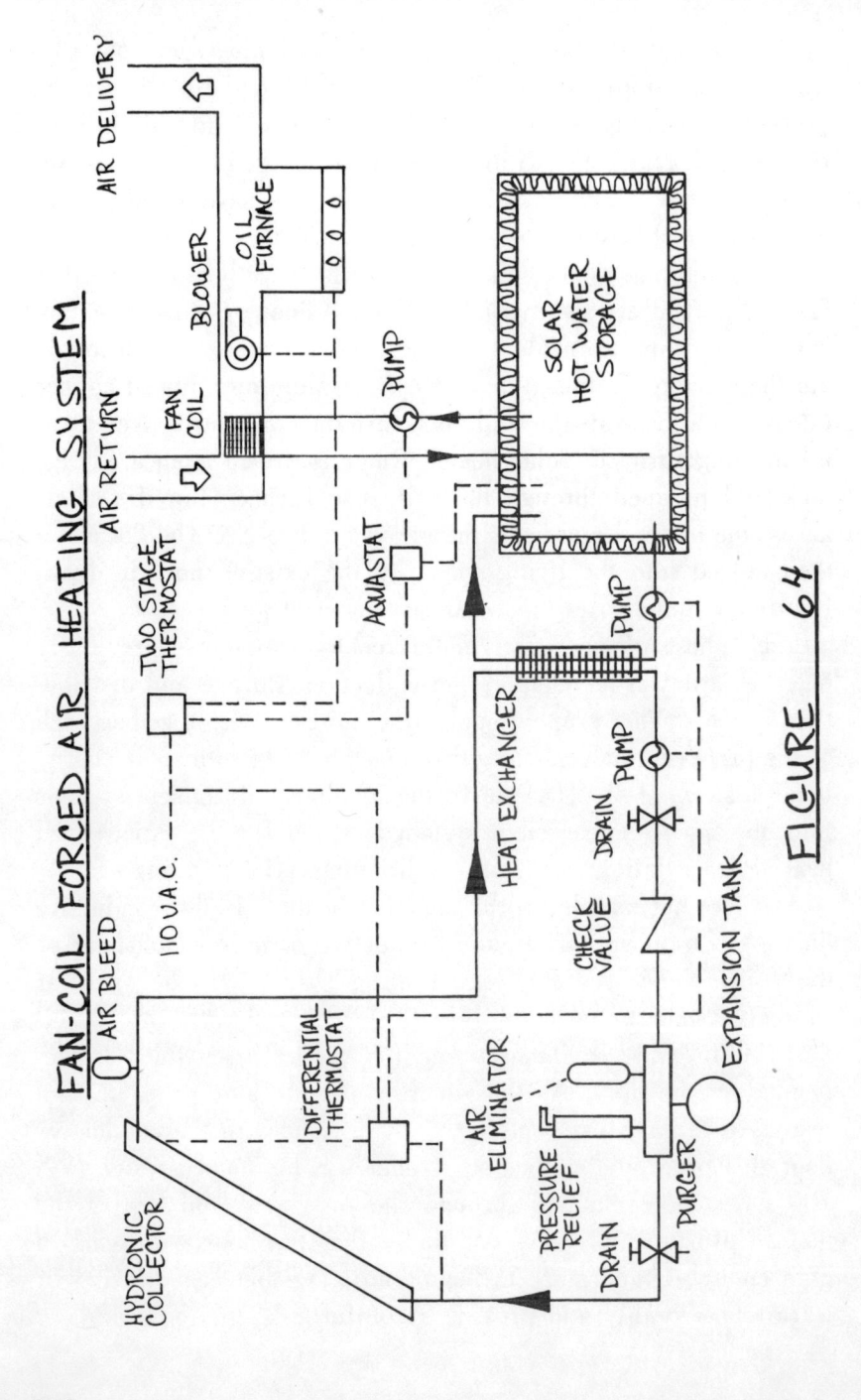

FAN-COIL FORCED AIR HEATING SYSTEM

FIGURE 64

140

6: Solar Systems

SOLAR CONDUCTIVE FLOOR

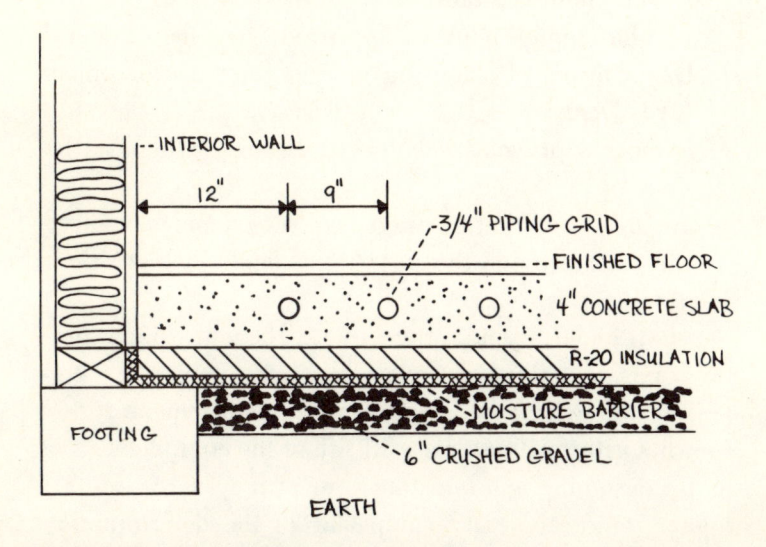

```
INTERIOR WALL
12"  9"
3/4" PIPING GRID
FINISHED FLOOR
4" CONCRETE SLAB
R-20 INSULATION
MOISTURE BARRIER
6" CRUSHED GRAVEL
FOOTING
EARTH
```

FIGURE 65

The conductive solar floor heating configuration is illustrated in Figure 65. Although limited in residential applications, its potential for meeting shop, garage and basement requirements is significant.

All of the active hydronic solar systems discussed in this chapter have been used effectively in the Northeast. In many instances, especially space heating, the initial costs of the systems are high. It is difficult to predict exactly how long it will take to recover an investment in solar equipment due to fuel savings when the future cost of fuel is unknown. It can be said with certainty, however, that solar domestic water heaters will recover their entire cost within at least ten years even at present fuel rates. Alternative systems such as the fan-coil forced air heating show tremendous potential as fuel oil prices exceed $1 per gallon and continue to rise unchecked.

FINAL CHECKLIST

If you are purchasing a manufactured system:
- —is your home suitable to capitalize on solar energy?
- —is the manufacturer approved by the Federal Department of Housing and Urban Development (H.U.D.)?
- —is there a proven track record of successful installations?
- —are there warranties or guarantees on performance?
- —who will provide the installation and what are their qualifications?
- —what Federal and/or State tax benefits, loan programs, and grants are available?
- —how extensive are the maintenance requirements?
- —what is the estimated useful life of the equipment?

If you are designing and building your own system:
- —are you technically competent in the determination of system components?
- —have you priced out all of the items which are necessary to complete the job?
- —is the overall system design efficient?
- —is the collector design efficient?
- —do you have the necessary trade skills to do plumbing, carpentry, sheet metal and electrical work?
- —how much maintenance will be required and will all items be accessible?
- —how will you integrate the conventional backup system?

Given the energy situation which exists today, it is difficult to look forward to a continuing reliance on limited supplies of high priced fossil fuels. The turn to solar power and energy self-reliance is not going to take place without a considerable change in the lifestyle and standard of living we currently enjoy. Follow-

6: Solar Systems

ing consideration of the solar alternative, many people begin to prepare to meet this challenge. The preparation does not take place without a great deal of time and devotion to the thorough analysis of both the problem and the potential solutions.

If you have committed yourself to the pursuit of solar energy as one small part of the solution, then your work has only begun. Determining how to put solar energy to use in your home has been the focus of this writing. The implementation of these concepts is now your responsibility.

Appendix A: Manufacturers of Solar Equipment

The following is an abbreviated listing of several of the more prominent solar equipment manufacturers, listed alphabetically by state. For a more detailed listing contact the National Solar Heating and Cooling Information Center, P.O. Box 1607, Rockville, Md. 20850, or call toll free: (800) 523-2929.

Fafco
138 Jefferson Drive
Menlo Park, CA 94025
(415) 321-6311

Piper Hydro Inc.
2895 E. LaPalma
Anaheim, CA 92806
(714) 630-4040

Raypack Inc.
31111 Agoura Road
Westlake Village, CA 91359
(213) 889-1500

American Heliothermal Corp.
2625 S. Santa Fe Drive
Denver, CO 80223
(303) 421-8111

International Solarthermics Corp.
P.O. Box 397
Nederland, CO 80466
(303) 258-3272

Solar Energy Research Corp.
701B S. Main Street
Longmont, CO 80501
(303) 772-8406

Solar Seven Industries Inc.
3323 Moline Street
Aurora, CO 80110
(303) 364-7277

Solaron Corp.
300 Galleria Tower
720 S. Colorado Blvd.
Denver, CO 80211
(303) 759-0101

145

Solar Energy for the Northeast

American Solar Heat Corp.
7 National Pl.
Danbury, CT 06810
(203) 792-0077

Sunworks Div. of Enthone Inc.
P.O. Box 1004
New Haven, CT 06508
(203) 934-6301

National Solar Corp.
P.O. Box C
330 Boston Post Road
Old Saybrook, CT 06475
(203) 388-0834

Sun-Ray Solar Equipment
2093 Boston Ave.
Bridgeport, CT 06610

Thomason Solar Homes Inc.
6802 Walker Mill Road S.E.
Washington, D.C. 20027
(301) 292-5122

CSI Solar Systems Div.
12400 49th Street
Clearwater, FL 33520
(813) 577-4228

General Energy Devices
1753 Ensley
Clearwater, FL 33516
(813) 586-3585

Solar Energy Products Inc.
1208 N.W. 8th Ave.
Gainesville, FL 32601
(904) 377-6527

A.O. Smith Corp.
Box 28
Kankakee, IL 60901

Lennox Industries Inc.
350 S. 12th Ave.
Marshalltown, IA 50158
(515) 754-4011

DuMont Industries
Main Street
Monmouth, ME 04259
(207) 933-4281

Acorn Structures Inc.
P.O. Box 250
Concord, MA 01742
(617) 369-4111

Daystar Corp.
90 Cambridge Street
Burlington, MA 01803
(617) 272-8460

Appendix A: Manufacturers of Solar Equipment

Contemporary Systems Inc.
68 Charlonne Street
Jaffrey, NH 03452
(603) 532-7972

Zomeworks Industries
P.O. Box 712
Albuquerque, NM 87103
(505) 242-5354

Bio-Energy Systems Inc.
Mountaindale Road
Spring Glen, NY 12483
(914) 434-7858

Grumman Energy Systems Div.
4175 Veterans Memorial Highway
Ronkonkoma, NY 11779
(516) 575-6205

Revere Copper & Brass Inc.
P.O. Box 151
Rome, NY 13440
(315) 338-2401

LOF Solar Energy Systems
1701 Broadway
Toledo, OH 43605
(419) 247-4355

Mor-Flo Industries Inc.
18450 S. Miles Road
Cleveland, OH 44128
(216) 663-7300

Owens-Illinois, Inc.
P.O. Box 1035
Toledo, OH 43666
(419) 242-6543

General Electric Company
P.O. Box 13601/Bldg #7
Philadelphia, PA 19101
(215) 962-2112

State Industries Inc.
Cumberland Street
Ashland City, TN 37015
(615) 792-4371

American Solar King Corp.
6801 New McGregor Highway
Waco, TX 76710
(817) 776-3860

Northrup Inc.
302 Nichols Drive
Hutchins, TX 75141
(241) 225-4291

Intertechnology Corp.
100 Main Street
Warrenton, VA 22186
(703) 347-7900

Appendix B: Additional Reading

The books listed below are outstanding sources for a more detailed look at specific topics in solar energy.

The Passive Solar Energy Book, Edward Mazria A.I.A. (Rodale Press, 1979). A detailed and well organized examination of the use of passive solar energy in building design.

Solar Energy: Fundamentals in Building Design, Bruce Anderson (McGraw Hill Book Company, 1977). A comprehensive review of research completed to date in solar energy applications.

Solar Heating Design, William Beckman, Sanford Klein and John Duffie (John Wiley and Sons, Inc., 1977). A thorough analysis of the physical parameters involved in the design of solar systems.

Solar Heating and Cooling, Jan Kreider and Frank Keith (McGraw Hill Book Company, 1975). An inspection of the fundamentals of heat transfer as applied to solar energy as well as an economic analysis of system performance.

The Solar Decision Book, Richard Montgomery (Dow Corning Corp., 1978). A complete look at all phases of the solar decision making process including a detailed review of all components involved in solar systems.

Appendix C: Glossary of Terms

Absorptance: The ratio of energy absorbed by a surface to total energy striking the surface, expressed as a percent.

Active Solar System: A system which requires additional energy (usually electric) to power pumps or fans which are necessary for the system's operation.

Air Type Collector: A solar collector which uses air as the heat transfer medium.

Ambient Temperature: Surrounding air temperature.

Angle of Incidence: Angle at which solar radiation strikes a surface.

ASHRAE: American Society of Heating, Refrigeration and Air Conditioning Engineers, Inc., 345 East 47th Street, New York, New York 10017.

Auxiliary System: A backup heating system utilizing conventional energy sources such as oil, electricity or natural gas to provide heat when the solar system can not.

Azimuth: The angle between true south and the direction in which collectors are facing.

Berm: A man-made hill of earth which may surround a building wall and increase its insulation value.

Black Body: A theoretically perfect absorber having an absorptance ratio of 1.0.

Solar Energy for the Northeast

Btu (British thermal unit): A unit of energy defined as the amount of heat required to raise one pound of water by one degree Fahrenheit.

Building design load: The amount of heat loss which occurs in a structure under the severest normal winter conditions.

Caulking: The use of a pliable sealing material to fill in cracks or crevices which may cause heat loss.

Clerestory: vertical window space located near the peak of the home to admit light.

Closed loop: A piping loop in which fluid is continually recirculated without being vented to the atmosphere.

Coefficient of heat transfer: The U-value or rate at which heat moves through a substance.

Collector: A device used to capture and transfer solar radiation.

Collector Angle: Angle between the collector plane and the horizontal surface of the earth.

Collector, linear focusing: A solar collector in which all energy is refocused and concentrated on a line, that line normally being a pipe through which fluid is circulated.

Collector, planar: A flat-plate solar collector which absorbs diffuse as well as direct radiation.

Collector, point focusing: A solar collector in which all energy is refocused and concentrated on a point.

Condensation: Droplets of water vapor which accumulate on the inside of a collector cover unless a desiccant is employed.

Conductance: The quantity of heat flowing through one square foot of a material in one hour when there is one degree Fahrenheit temperature difference between the two surfaces.

Conduction: One of the three methods of heat transfer in which

Appendix C: Glossary of Terms

heat moves through a solid mass by molecular excitation of adjacent molecules.

Conductivity (k): The quantity of heat flowing through one square foot of a material which is one inch thick in one hour when there is a temperature difference of one degree Fahrenheit between its surfaces.

Convection: One of the three methods of heat transfer in which heat is moved in a body of liquid or air.

Counterflow heat exchanger: A device in which two fluids flow in opposite directions allowing for the transfer of heat from one fluid to the other.

Coverplate: A transparent sheet of glass or plastic mounted above the collector plate.

Degree Day: A unit of measurement used in heat loss calculations which shows degrees difference between 65°F and the day's mean outdoor temperature.

Delta T: A difference in temperature.

Density: Weight per unit usually expressed in pounds per cubic foot.

Desiccant: A material which characteristically absorbs moisture.

Differential Thermostat: A device which measures two inputs in order to effect a decision. In solar systems, the two inputs are normally collector plate temperature and storage temperature.

Diffuse radiation: Scattered, non-parallel rays of solar energy resulting from dust, moisture and various inconsistencies in the earth's atmosphere.

Direct Radiation: Rays of solar energy which have travelled in a straight path from the sun.

Drumwall: A heat storage mass utilizing stacks of waterfilled 55 gallon drums painted black.

Efficiency: Ratio of energy striking a solar collector to energy output of the collector.

Emissivity: The measure of a surface's tendency to emit thermal radiation.

Energy: The ability to do work.

Eutectic Salts: Salt compounds used to store heat in the change of phase from solid to liquid and release that heat as they re-solidify.

Expansion Tank: A device used to take up overflow as a heated fluid expands.

Flash Point: The temperature at which fluid vapors will flame if an ignition source is present.

Flat-plate Collector: See collector, planar.

Galvanic Corrosion: Corrosive action which arises due to the placement of dissimilar metals in contact with each other.

Glazing: The transparent glass or plastic material which is above the collector plate.

Greenhouse effect: The characteristic tendency of certain materials to be transparent to short wave radiation and opaque to longwave radiation.

Headers: Primary passages through which the fluid enters or leaves the collectors.

Heat Capacity: Amount of heat that can be stored in a specific material and raise the temperature by 1°F.

Heat Exchanger: A device which transfers heat from one substance to another without mixing the two.

Heat Pump: A device which uses a compressible refrigerant to transfer heat from one source to another.

Heat Transfer Media: Air or liquid which is heated and used to move energy from one place to another.

Appendix C: Glossary of Terms

Heat Transmission: The rate at which heat is moved through a solid.

Hysteresis: Factory setting on differential controls to prevent unit from flickering on and off when temperature difference is small.

Incidence: Thermal radiation falling on a surface.

Infiltration: The uncontrolled entry of air into a building through cracks and crevices.

Infrared Radiation: Electromagnetic radiation commonly experienced as heat.

Inhibitors: Additives to storage water to prevent algae, and corrosion.

Insolation: The total amount of incoming solar radiation striking a surface which includes direct, diffuse and reflected components and is usually expressed in Btus per square foot per hour.

Insulation: Materials used to prevent heat loss or heat gain usually by employing small dead air spaces which inhibit convection and conduction.

Isogonic chart: Shows magnetic compass deviations from true north.

Latent Heat: A change in heat content that occurs without a change in temperature.

Latitude: The angular distance north or south of the equator measured in degrees of arc.

Log Mean Temperature Difference: A complex mathematical function used to predict fluid temperatures in a heat exchanger.

Nocturnal Cooling: The cooling of a structure to the night sky.

Opaque: Impenetrable by light.

153

Solar Energy for the Northeast

Passive Solar System: A system which requires no outside energy for pumps or fans to effect its operation.

Payback Period: The amount of time required to recover the initial investment due to fuel savings.

Phase changing materials: Certain materials used in latent heat storage techniques.

Radiation: One of the three methods of heat transfer in which energy is transmitted through space by electromagnetic waves.

Reflectance: The ratio of the amount of light reflected by a surface to the amount incident.

Retro-fitting: The installation of solar equipment on buildings which were not designed for that purpose.

R-value: Coefficient of resistance to heat transfer.

Selective Surface: A surface with a coating that has a high absorptance of incoming radiation but a low emittance of infrared radiation.

Sensible Heat: Heat that results in a temperature change.

Specific Heat (Cp): The number of Btus required to raise the temperature of one pound of a substance 1°F.

Stagnation: A condition that results when heat transfer fluid is not flowing in a solar system.

Stratification: The occurrence of layers of different temperatures in a material where the top layers are warmer than the bottom.

Thermal Mass: The amount of potential heat storage capacity in a given volume.

Thermo-siphon: A natural convective system in which no pump is required to cause circulation.

Transmittance: The ratio of energy passing through a surface to the energy incident.

Appendix C: Glossary of Terms

U-value: Coefficient of heat transfer, the number of Btus that move through one square foot of a material during one hour when a 1°F temperature difference is present.

Vapor Barrier: A construction component used to prevent the passage of condensation from one surface to another.

Viscosity: The resistance of a fluid to movement.

Wavelength: The length in microns between the start and finish of an energy pulse.

Index

active system, 26
altitude, 17
ASHRAE (American Society of Heating, Ventilating and Air Conditioning Engineers), 11
azimuth, 39

Baer, Steve, 126
bank loans, 36
British themal unit (Btu), 18

coatings, absorber plate, 52
coatings, properties of, 55
collector, absorber plate, 43-51
collector, flat-plate, 29
collector, frame, 64
collector, general, 43
collector, linear, 26
collector, passive air, 124
collector, planar, 28
collector, point focusing, 26
collector, roof integrated, 65
collectors, types, 26
conduction, 108
conductive heating, in floor, 140
convection, 108
continuous air circulation, 9

"delta T", 66
design temperature, 11
dessicants, 59
differential temperature, 73
distribution systems, air, 110
distribution systems, hydronic, 112
Drumwall, 126
domestic hot water, 32
domestic hot water system, antifreeze, 134
domestic hot water system, draindown, 132
dryers, 33

efficiency curves, 66
efficiency, for several applications, 70
efficiency, variation in absorber plate, 46-48
elastomeric pillows, 96
electromagnetic spectrum, 54
energy offices, state and federal, 37
EPDM rubber, 65

equations, for calculating performance, 74
ethylene glycol, 77
extruded aluminum, 64

financing, 35, 38
flammability, in fluids, 76
floor plenum, 111
flow patterns, 50
freezing point, in fluids, 76
fuel cost projection, 35

gaskets, 65
Glaubers salt, 82
glazing systems, 57
glazings, dual vs. single, 58
greenhouses, 33, 124

Hay, Harold, 126
heat exchangers, 91
heating degree day, 12
heat pumps, 113
heat transfer, methods of, 108
heat transfer fluids, 75
heat transfer fluids, properties of, 78-81
Heliolite, Combustion Engineering Corp., 58
H.U.D., 30

inclination, 39
insolation values, 18
insulating systems, 61
insulation, properties of, 63

kinetic energy, 82

latitude, 42
Löf, Dr. George, 101

magnetic deviation, 41
manufacturers, 145-147
mass flow rate, 72
National Energy Act, 38

oil, aromatic, 77
oil, hydrocarbon, 79
oil, parrafinic, 77
oil, silicone, 77

passive solar design, 5
pool blankets, 139
pool heating, 33, 137

Solar Energy for the Northeast

potential energy, 82
property tax assessment, 37
propylene glycol, 77

radiant floor heating, 113
radiation, 109
radiation, direct, 27
radiation, ratio of diffuse to total, 28
radiation, scattered, 27
radiation, through collector glazings, 56
reflectors, 18
rock bed storage, 98-101
roll-bond absorber plate, 49
"r" value, 62

sales tax, 37
S.E.I.A. (Solar Energy Industries Assoc.), 67
selective surface, 54
skin temperature, effect of, 79
sky cover, average, 15
Skytherm, 126
solarium, 126
space heating, 32
specific heat, 72
storage, 82-89
storage, alternative locations, 84
storage, calculating Btu output from, 102
storage, calculating heat loss from, 104
storage, domestic hot water, 88
storage, in attic, 96

storage, latent heat, 82
storage, sensible heat, 82
storage, space heating, 94
storage, Thomason System, 100
storage, volumes required, 84
structural integrity, 30
Sunadex, ASG Industries, 58
sunshine, hours of, 12
systems, active air, 129
systems, active hydronic, 131
systems, passive air, 120
systems, passive hydronic, 126
system, components, 136

tanks, 97
tax credits, 38
temperature, range, 51
thermal conductivity, 45
thermal curtain, 122
thermal expansion and contraction, 59
thermal isolation, 52
thermal resistance, 45
thermopane, 6
thermo-siphon, 128
Thomason, Dr. Harry, 99
toxicity, in fluids, 75
Trombe wall, 123

wetted surface, 46
window box, 122
wood burning, 9